Interstellar dust grain: diameter 4×10^{-5} inch

Blue light wavelength: 1.9×10^{-5} inch

Bacterium: diameter 4×10^{-5} inch

Black hole: diameter 40 miles

Large moon crater: diameter 120 miles

Largest asteroid: diameter 620 miles

Mars: diameter 4,217 miles

White dwarf: diameter 5,000 miles

Venus: diameter 7,521 miles

SPACEFARERS

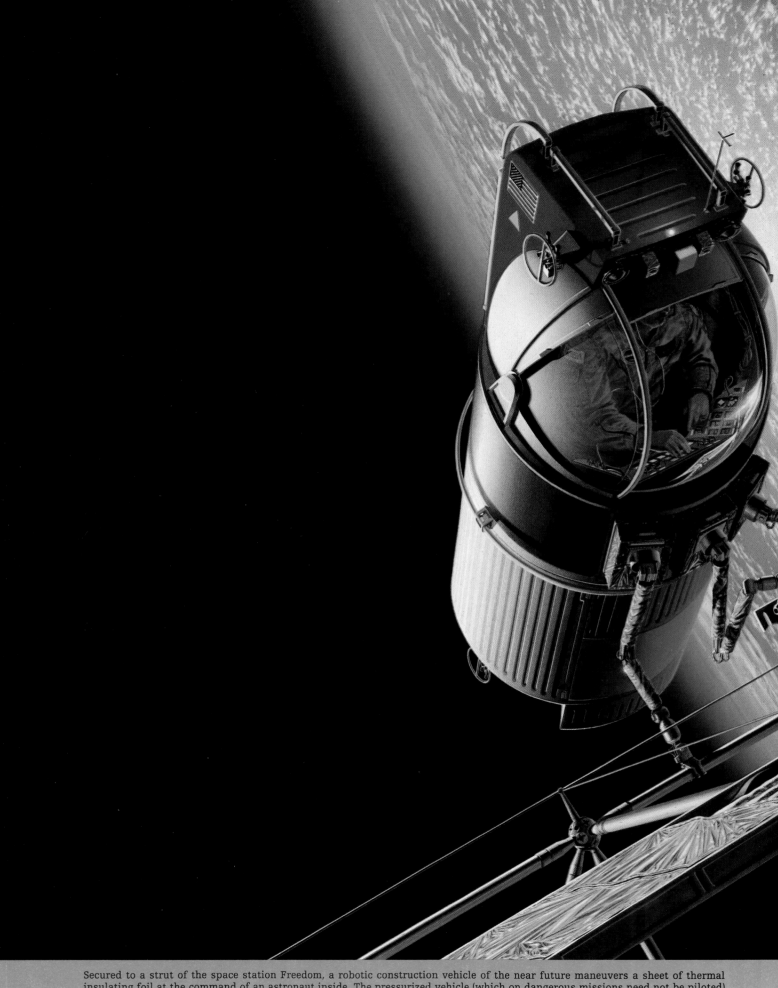

Secured to a strut of the space station Freedom, a robotic construction vehicle of the near future maneuvers a sheet of thermal insulating foil at the command of an astronaut inside. The pressurized vehicle (which on dangerous missions need not be piloted)

will be able to build large structures as well as perform delicate microelectronic repairs. Computers, communications equipment, lights, and cameras will be housed in its upper section; the lower portion will hold life-support and electrical systems.

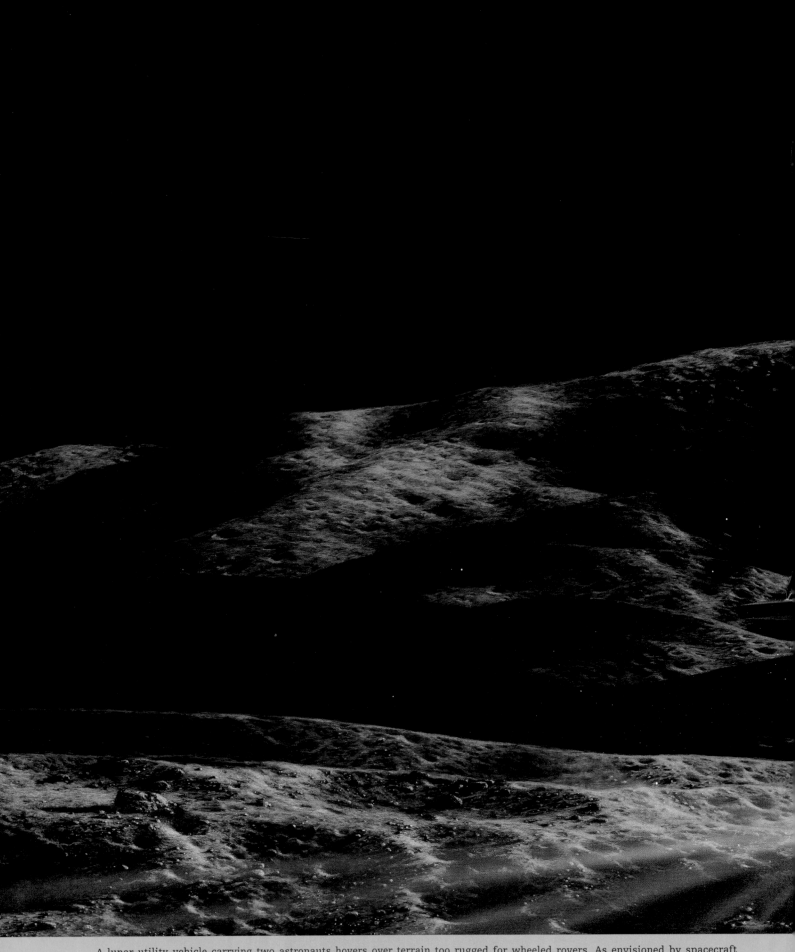

A lunar utility vehicle carrying two astronauts hovers over terrain too rugged for wheeled rovers. As envisioned by spacecraft designers, the six-engine craft will be able to take off and land vertically, ferrying five tons of cargo mounted under a quartet of fuel

tanks. Using communications equipment powered by solar panels and fuel cells, astronauts will remain in contact with their main Moon base and mission controllers on Earth while collecting data or making repairs at remote scientific sites.

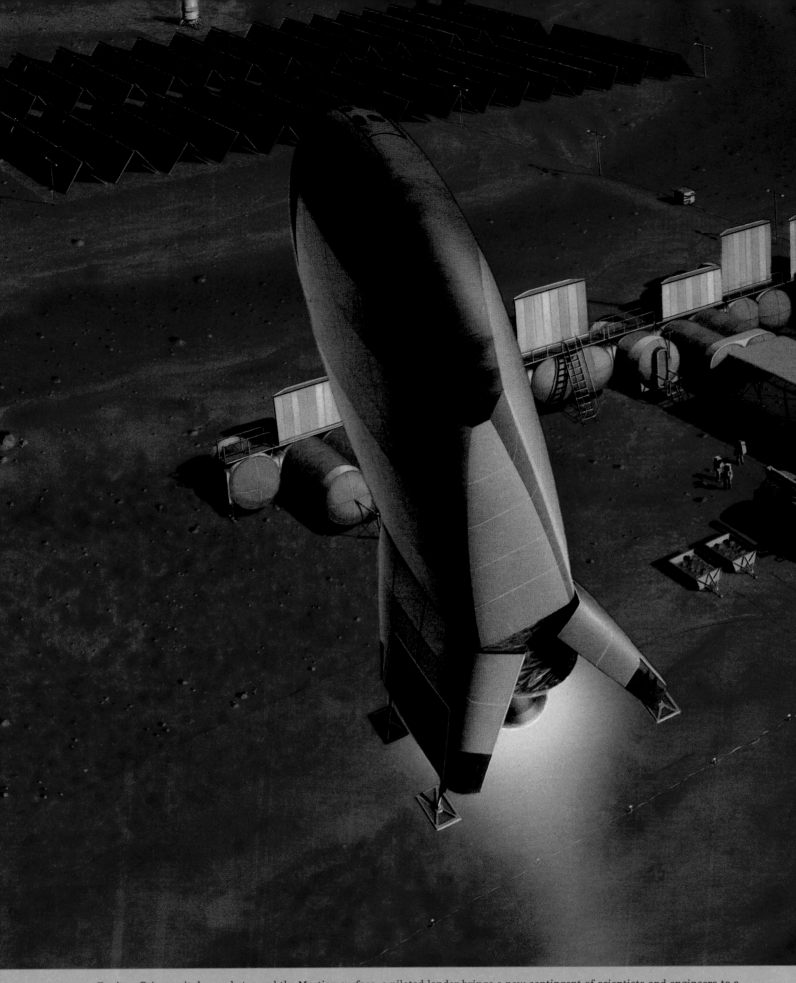

Engines firing as it descends toward the Martian surface, a piloted lander brings a new contingent of scientists and engineers to a research colony, where as many as thirty reside on a rotating basis. The complex—designed to provide as normal a life as possible

in the hostile environment—includes habitat, recreation, and scientific modules, as well as facilities for nuclear power and oxygen extraction. Around the colony, rovers hauling personnel and equipment leave telltale signs of human activity in the dust.

Hurtling past Pluto and its moon Charon, a starship powered by enormous ramjet engines plunges into deep space. Six-thousand-mile-long cones made of magnetically connected field generators scoop hydrogen atoms from the interstellar medium and

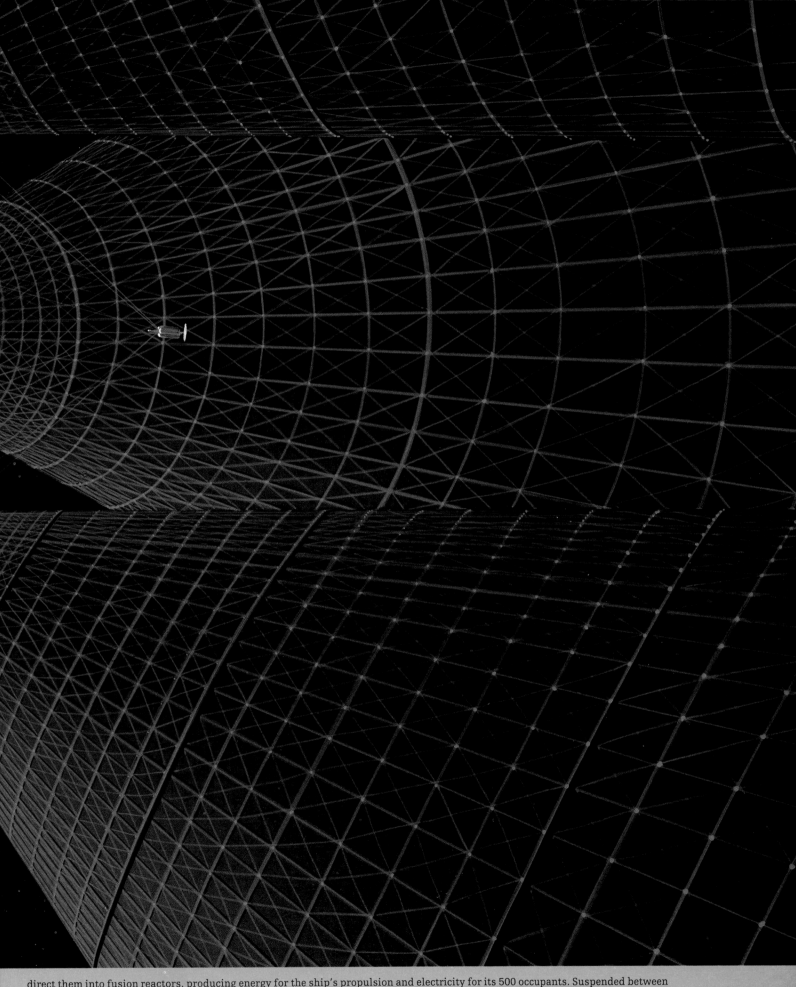

direct them into fusion reactors, producing energy for the ship's propulsion and electricity for its 500 occupants. Suspended between the cones are living quarters to house several generations of human beings as they search the galaxy for new worlds.

TIME LIFE ®

This volume is one of a series that
examines the universe in all its aspects,
from its beginnings in the Big Bang to the
promise of space exploration.

VOYAGE THROUGH THE UNIVERSE

SPACEFARERS

BY THE EDITORS OF TIME-LIFE BOOKS
ALEXANDRIA, VIRGINIA

CONTENTS

1/Earth Orbit

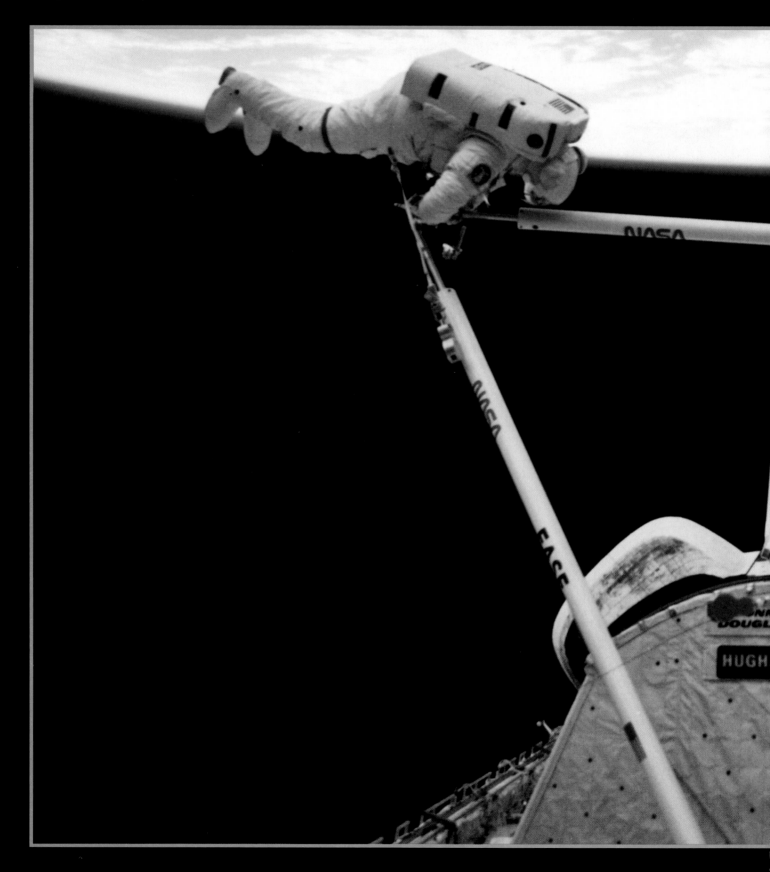

Floating 250 miles high in Earth orbit, shuttle astronauts practice assembling interlocking struts. When the U.S. space station is built in the 1990s, the parts will form a sprawling latticework frame holding living quarters and laboratories.

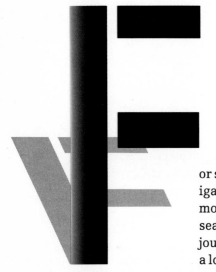or several human lifetimes, the huge spacecraft has navigated the unmarked lanes of the universe. Alone and moving in utter silence, it has coursed through a black sea lit by a billion glittering stars. Now the intragalactic journey is ending at last. Ahead, the voyagers gaze upon a long-anticipated sight: an unfamiliar planet, caught in a gauzy cocoon of clouds. Behind them, more than 100 trillion miles away, is where the expedition began—in the orbit of Earth, third planet from the midsize star called the Sun. Early in the twenty-first century, telescopes on Earth's moon revealed that a planet in the Delta Pavonis system might have the correct atmospheric chemistry to support human life. And it was attainable: With recent improvements in pulse fusion rocketry, a starship could travel at a significant fraction of the speed of light, allowing it to reach nearby stars within the reasonable timespan of two centuries or so.

Inevitably, the old wanderlust of the species asserted itself. Aboard a vast interstellar ark, 10,000 explorers set forth to seek a second home for humankind. As the craft flashed through the comet-strewn Oort cloud that marked the outer boundaries of the Solar System, the receding Sun took its place among the background stars. Within thirty years, the enormous vessel achieved its target velocity of one-tenth the speed of light. For the next century and a half, it coasted. Not until the last decade of the journey did the ship gradually slow, braking on its microthin boron sail.

By then, the sense of voyaging had been obscured by the rhythms of life. On board the ship, which spun on its axis to create artificial gravity, men and women married and bore children; the oldest travelers passed away. Work, families, and hobbies occupied the crew as they would have back on Earth. Engineers monitored the recycling systems that circulated water and oxygen through the spaceship complex. Hydroponics farmers nurtured dense plots of beans and rice. The craft had its own shops, a school, a library, a hospital, and a social center. Only the communications specialists, collecting increasingly outdated news from Earth, and the navigators, peering ahead to their goal, were clearly conscious of the journey's beginning and end. For the others, as time passed, Earth became little more than a treasured memory handed down like old china from parent to child.

But now, young and old suddenly remember who they are—or rather, who their ancestors were and what they were seeking. A band of scouts starts preparations to leave the ship in the aerospace plane stored in the cargo hold.

The great gamble approaches its climax. Have they made their trip for nothing? Or will they soon experience the terrible freedom of a new sky? . . .

THE FIRST STEP

. . . As the twentieth century draws to a close, interstellar exploration remains the heady stuff of fiction—a prospect far grander than anything that humans have yet attempted. But the foundations for travel across the cosmos are being laid even now. Around the world, experts in space science are planning for a lunar outpost and a piloted mission to Mars within the next few decades. Closer at hand are permanently staffed bases in orbit, descendants of the space stations of the 1970s and 1980s. In fact, it is this next wave of stations that may open the way for future voyages into deep space. They will, for example, provide scientists with opportunities to study the effects of weightlessness on the human body over long periods of time. And they should also allow researchers to test different systems for providing spacefarers with the food, water, and fresh air needed for survival during protracted spaceflights.

Perhaps more important, orbiting space stations may one day serve as launch sites for missions requiring spacecraft too complex and delicate to boost from Earth. Objects orbiting the planet are essentially weightless, balancing centrifugal force against gravity. Many engineers believe that the way to send a rocket to Mars, or to points beyond, would be to ferry the parts for the craft to an orbiting space station, where the ship would be assembled and fueled. At such a station, engineers could test the ship's flight readiness in a vacuum, thus facing far less risk of a total mission failure. When building materials and fuel become available from extraterrestrial sources—say, the Moon or the asteroids—transportation costs should drop, since those items will no longer have to be lifted out of Earth's gravitational field.

The idea of humans occupying a base orbiting Earth, like so many aspects of space travel, had its genesis in fiction. In 1869, a short, whimsical tale appeared in the *Atlantic Monthly* magazine. The story described the exploits of thirty-seven New Englanders accidentally thrust into orbit aboard a 200-foot brick sphere launched by the force of two enormous flywheels rotating in opposite directions. Once over their initial shock, the first space-station tenants established an idyllic society. They grew a variety of useful crops, and in time found that they had less and less interest in the affairs of Earth.

The author of this fable, entitled, appropriately, "The Brick Moon," was not a science-fiction buff. Rather, he was a Boston clergyman and writer with a dry wit and a zeal for social reform, perhaps inherited from his great-uncle, the Revolutionary War hero Nathan Hale. Although he intended his tale to be social commentary, Edward Everett Hale raised concerns that would confront engineers and designers many decades later as they considered the challenges of spaceflight—construction, propulsion, and food supply. For example, although modern spacefarers have in fact managed to build and launch craft into orbit, they have yet to find a way to grow anything except one kind of weed through an entire life cycle. "The Brick Moon" enjoyed a brief popularity, then faded into obscurity until the first Sputnik satellite was rocketed into orbit in 1957, sending reporters scurrying for relevant lore.

In the meantime, futurists with more practical goals began to explore the design and function of orbiting colonies. Among these space advocates was Konstantin Tsiolkovsky, a Russian mathematician and teacher and a progenitor of modern rocket science. Born in 1857, Tsiolkovsky was the son of a forester whom he described as an impractical dreamer. Yet the boy seemingly inherited his father's idealistic tendencies. In part because an early illness had left him nearly deaf, Tsiolkovsky led a solitary childhood reading and fantasizing about space travel. Later, as an adult, he recalled that he had always felt a desire to reach out toward the Sun, and had yearned for "a realm devoid of gravity, where one could move unhampered anywhere, freer than a bird in flight."

Perhaps motivated by such passions, Tsiolkovsky in 1920 published *Beyond the Planet Earth,* a novel describing life on a spindle-shaped craft in permanent orbit around Earth. On board was a multinational crew (the Russian visionary also anticipated the era of global politics) who built greenhouses aloft and harnessed the Sun's energy for power. So attractive was orbital existence that thousands of volunteers followed the first crew into space and established a giant network of colonies. Tsiolkovsky located the colonies in an orbit some 22,000 miles above Earth, where they traveled at about 7,000 miles per hour, circling the planet in exactly the time it takes Earth to rotate once on its axis. Today's fleet of communications satellites make a similar geosynchronous orbit, as it is called, around the globe.

Tsiolkovsky's book appeared at the start of an era in which space travel seemed to grip the public's imagination. In 1930, an eighteen-year-old German engineering student named Wernher von Braun—who would go on to develop the V-2 rocket that was used in World War II—read about a lunar voyage in a pulp science-fiction magazine and began to dream of leaving the Earth's surface. After the defeat of Germany, the American military lured von Braun to the United States, where, in the early 1950s, he wrote his own fictional tale about an inhabited space wheel. Published in a *Collier's* magazine series on space travel, the story was bolstered by von Braun's expert knowledge of the difficulties of escaping Earth's gravity. An expedition might more easily depart for the Moon and Mars from a space station, he explained.

Von Braun also anticipated one of the prerequisites for a viable base in orbit, a freighter to transport construction parts up from the Earth's surface. He described the ship in detail: The cone-shaped freighter would have two huge wings mounted on its stern and would be capable of carrying a crew and a thirty-six-ton payload into orbit. When its mission was over, it would fire braking rockets to take it out of orbit and back into the atmosphere. As its wings caught the atmosphere, the craft would begin a gliderlike descent to a runway landing. Later, the freighter could be refurbished for use on another mission. Von Braun, who died in 1977, did not live to see his dream come true. But in the 1980s, both the United States and the Soviet Union launched winged spaceships that shuttled to and from orbit in much the way he had described.

Another space enthusiast inspired by teenage reading of science fiction was Arthur C. Clarke, who grew up in the coastal town of Minehead, England. In time, he would turn his own visions into both fiction and fact. While serving in the Royal Air Force during World War II, shortly before he acquired a degree in physics and mathematics from Kings College, London, Clarke began to write about space and space travel. In 1945, more than a decade before they became reality, he published a technical article about communications satellites in geosynchronous orbit. Then, in 1951, he published a nonfiction book called *The Exploration of Space* that proved even more prescient.

Clarke devoted a full chapter of his book to space stations, which he predicted would be assembled in orbit by spacesuit-clad crews working with materials ferried up by spaceships. Once complete, Clarke wrote, the stations would serve as research bases for studies in astronomy, meteorology, and biology. They would also be useful for refueling other spacecraft and as radio-communications relay points. Looking farther into the future, Clarke anticipated that the stations might "eventually become miles in extent—in fact, veritable cities in space."

Today, many scientists acknowledge that their own ideas of extraterrestrial travel were heavily influenced by visionaries such as Clarke and Tsiolkovsky. In the late 1950s and early 1960s, the first generation of space scientists began to put these ideas into practice.

A CROP OF VETERANS

Cosmonaut Yuri Gagarin's 108-minute ride into space aboard the Soviet craft *Vostok 1* in 1961 was a far cry from the sophisticated voyages of Clarke's fiction. Still, it marked humanity's first steps into space—and the onset of the space race. By the end of 1961, two Ameri-

HOUSES IN THE SKY

As seen at right and on the following pages, the notion of living in space has evolved from purest fantasy to familiar fact in just over a century.

1869 American clergyman Edward E. Hale *(near right)* described an artificial satellite in a story titled "The Brick Moon." Built with 12 million bricks, his edifice *(far right)* would house thirty-seven inhabitants.

cans and another Soviet had made the trip, the cosmonaut for a full day. By the end of the decade, dozens of men and one woman were space veterans; some had gone as far as the Moon.

For the first few years, mission planners kept flights intentionally short to protect astronauts from the unknown dangers of the alien environment. Until December 1965, no Soviet or American flight lasted more than eight days. Then Frank Borman and James Lovell, circling Earth aboard *Gemini 7*, set a new record: thirteen days, eighteen hours, and fifty-five minutes aloft. The Soviets passed this milestone four and a half years later, when two cosmonauts aboard *Soyuz 9* remained in space for eighteen days in June 1970.

Scientists in both countries knew that interplanetary travel would come to pass only if they could determine that people could live in space for months or years at a time. Cramped one- or two-person rocket ships clearly were not the proper test bed; instead, space planners had to build orbiting habitats far larger and more complex than any spacecraft yet attempted.

The Soviet Union was first off the launch pad. On April 19, 1971, a powerful Proton rocket boosted the world's first space station, *Salyut 1*, into orbit. The twenty-ton station, named "salute" in honor of the tenth anniversary of Gagarin's flight, was a modified spaceship, a metal cylinder some sixty-five feet long and thirteen feet in diameter that sported winglike solar panels. Inside, a trio of pressurized compartments would house a crew of three.

Despite its simplicity, *Salyut 1* represented a radical step for the Soviet space program. Circling in low Earth orbit, or LEO, close to 250 miles high, the ship was within easy reach of spacecraft carrying crew or cargo. For the first time, people would really be living in space, not just rocketing around the Earth breaking records for number of orbits. Two months after *Salyut 1*'s launch, three cosmonauts aboard *Soyuz 11* blasted away from Earth and maneuvered their spacecraft around the globe until they spotted the empty station gleaming ahead of them. Adjusting their trajectory with the delicate touch of attitude jets, the crew rendezvoused with the station, docked at the

single port, and clambered aboard. For twenty-three days, the cosmonauts reveled in their roles as the first true space dwellers. Back on Earth, Soviet citizens sat glued to television sets nightly, watching broadcasts of the Salyut team's midair stunts in the weightless environment. The watchers also marveled as the spacemen grew beards and toasted a crewmate's birthday with fruit juice, as if to say that life aloft was not all that different from life on Earth, even if the juice had to be sipped through tubes.

But the Soviet Union's euphoria was shattered at the mission's end. When the returning Soyuz capsule finally settled gently in a remote Kazakhstan wheat field and the ground crew opened the hatch, they found the three cosmonauts strapped in their seats, dead. There were rumors that some unknown factors in space had killed them; perhaps humans were not meant to live beyond Earth's surface. Eventually, however, engineers determined that a crucial valve connected to the cabin had loosened, depressurizing the spacecraft during reentry. Because the three were not wearing spacesuits—deemed too bulky for the small Soyuz—they died when their air bled away.

The tragedy set the Soviet space program back several years. During the interim, the United States began to develop Skylab, its own version of a space station. Built from the remnants of an Apollo Moon rocket, Skylab was a bulky craft with solar panels mounted like helicopter blades on its nose. With a 10,000-cubic-foot interior unevenly divided into two compartments, it offered its crew about the volume of a three-bedroom house—roughly triple the space inside Salyut. The spacious upper deck, which took up about three-quarters of the interior room, was used by the astronauts for most daytime activities; the snug lower deck was where the crew slept.

Skylab was launched atop a Saturn booster rocket on May 14, 1973. Eleven days later, its first inhabitants arrived aboard their Apollo command ship, docked at the station's bow, and entered for a twenty-

1895 Russian mathematician Konstantin Tsiolkovsky *(below, left)* proposed to escape Earth's gravity with a 22,500-mile tower. At such a height, gravitational pull would be so weak that the centrifugal force of Earth's rotation might be able to slingshot a craft into space.

1926 Hugo Gernsback *(below, right),* an American engineer and entrepreneur, founded the world's first science-fiction magazine, *Amazing Stories.* A mix of fact and prophecy, the magazine inspired writers and scientists alike to pursue the dream of spaceflight.

eight-day residency. Eventually, Skylab would host two more teams, the last of which would remain on board for eighty-four days. Never before had human beings spent so much time away from Earth. As Captain Alan Bean, the commander of the second crew, remarked before departing on his own fifty-nine-day stint in space: "We really don't know what's going to happen to the guys in the long term, and finding out is probably, in my mind, the single most important thing we've got to do in Skylab."

Despite this solemn pronouncement, the Skylab crews managed to have fun aboard the station, soaring about the upper compartment and performing floating acrobatics. "We were tickled to death!" exclaimed Charles "Pete" Conrad, Jr., commander of the first crew. "We never went anywhere straight; we always did a somersault or a flip on the way." The astronauts also enjoyed experimenting with various objects, many of which behaved quite differently in space than on Earth. They discovered, for example, that when playing darts they had to shoot the Velcro-tipped projectiles absolutely straight, rather than aim a bit high as required on Earth, where gravity affects the trajectory. The crew also noticed that in the absence of gravity, water was shaped by surface tension. When the astronauts used a syringe to squeeze a water drop into the air, the liquid would immediately form a perfect tiny globe. The floating sphere could be pulled and stretched into different shapes—a ribbon, a rope, or a flat plane—as if it were made of modeling clay.

In addition to conducting such playful scientific demonstrations, the Skylab team performed a wide range of research tasks. One of the most important jobs they had was that of keeping meticulous accounts of their own bodily functions. They reported every mouthful of food they consumed, drew daily blood samples, and hooked up machines to record their brain waves during sleep. The ground crew also monitored the astronauts, eavesdropping shamelessly to gauge the psychological effects of the experience. They gathered masses of data, creating a base of knowledge to help engineers design the next space stations and guide others sojourning in orbit.

1951 British writer Arthur C. Clarke *(below, left)* published space-station plans showing remarkable scientific foresight. As pictured here *(far left)*, the station incorporates a pressurized crew chamber *(foreground)*, a radio antenna *(upper left)*, and a parabolic structure to focus sunlight for energy.

1952 Wernher von Braun *(center)*, an eminent German-American engineer, wrote in *Collier's* magazine about a space station that rotated to simulate terrestrial gravity.

1969 Physicist Gerard O'Neill *(right)* began work on plans for a space colony populated by 10,000 inhabitants. It would occupy a gravitationally stable spot between the Earth and Moon known as L-5.

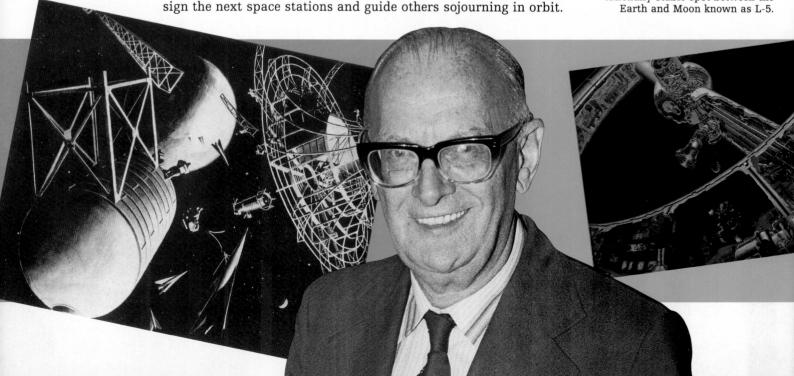

Skylab itself, however, was not meant to be a permanent home in space; mission planners knew its systems would fail eventually and gravity would pull it down. After the third crew departed in February 1974, the station remained empty. In July 1979, Skylab lost the battle with gravity, slipping back toward Earth and tumbling through the atmosphere in a shower of blazing debris, scattering pieces across the Indian Ocean and Australia.

The Soviets, meanwhile, had resumed their space-station program, and in 1977 two cosmonauts took up life aboard *Salyut 6,* first of the new generation of Salyut stations. Both *Salyut 6* and *Salyut 7,* which followed it in 1982, were a significant improvement in design over earlier American and Soviet stations. Attached to the new models was a second docking port, which made for more efficient resupply missions. The resident crew could remain on board to help unload cargo rather than having to move their own Soyuz vehicle to make room for the supply ship to dock. More important, as long as no cosmonauts were scheduled to go up to or return from Salyut, mission planners could also dispatch supplies on an unpiloted cargo vessel, thereby cutting costs. By March 1978, a team of Soviets had spent ninety-six days in orbit. Another *Salyut 6* crew stayed aloft for 185 days, and in 1984, three cosmonauts spent 237 days living on *Salyut 7*—well over half a year of spaceflight.

Then, in February 1986, the Soviets launched Mir, the flagship of the third generation. With six docking ports, Mir could accept various scientific and supply modules, giving the cosmonauts more room for their activities. When experiments were completed, the data and samples would return to Earth on a Soyuz ferry ship. Because of its comfort and versatility, the new station was a great success, and late in 1988 a Soviet crew returned from Mir after spending a record 366 days in space.

With the 1988 mission, Soviet crews demonstrated that people could stay alive and relatively healthy in space for close to the duration of an interplanetary voyage. Still, the alien conditions of space did take some toll on bodies designed for life on Earth.

THE EFFECTS OF WEIGHTLESSNESS

The main concern facing those who might spend months or years on a space station or Mars-bound ship is the potentially debilitating effect of weightlessness—or more accurately, microgravity, since the mass of a space station produces about one ten-thousandth the force of Earth's gravity. More than half of the travelers newly arrived in orbit complain of a feeling of malaise akin to seasickness. The discomfort, which seems to result from inner-ear disturbance, usually passes within a few days.

Another early effect, however, is longer lasting. Because the full force of Earth's gravity is no longer present to draw blood and other body fluids into the lower limbs, the fluids migrate toward the chest and head, giving crew members puffy faces. The fluids accumulating in the chest cause the body to register an excess; in response, the body sheds, through the kidneys, between 10 and 20 percent of its total fluid volume. Astronauts apparently suffer no ill effects from the loss while in space, but on return to Earth the deficit can make them weak and lightheaded. For interplanetary explorers, with perhaps only a few days or weeks to spend on the surface of a new world before returning home, the time lost to such symptoms could endanger the mission.

In an effort to counteract these ills, space physiologists have devised several procedures. Crews aboard Salyut and Mir, for example, have spent short periods each day inside a negative body pressure suit that pulls on the lower extremities like a vacuum cleaner to draw blood into the legs. Then, before reentry, they don another garment, known as an antigravity suit, that prevents too much blood from rushing downward once they reenter Earth's gravity field. In addition, returning spacefarers drink extra water with salt before touchdown to increase their fluid volume and prevent fainting.

Muscles also suffer in microgravity because most activities simply require less effort. Soviet researchers found that cosmonauts who spent just one month in space lost anywhere from 10 to 20 percent of the muscle strength in their arms and legs. (One

1973 The United States launched its first space station, Skylab *(far left)*. Conceived and built to operate for nine months, the station hosted three crews who performed a wide range of scientific experiments.

1986 Soviet space station Mir *(near left)* went into orbit. Unlike the earlier Salyut series, Mir featured six docking ports, to accommodate a variety of spacecraft, as well as additional living quarters and laboratories.

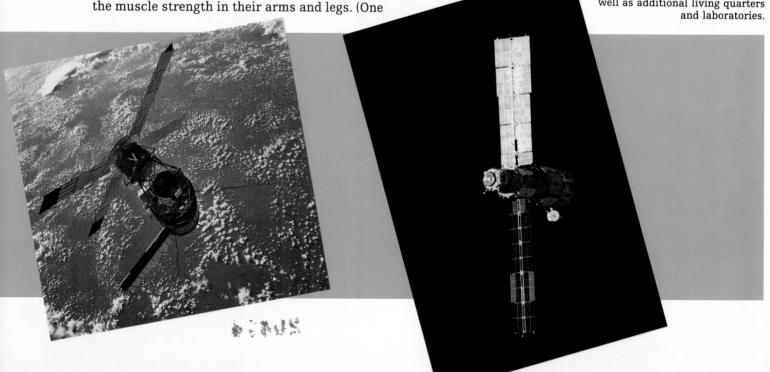

activity that has proved more difficult in space is the act of bending over: The Skylab crew reported that putting on shoes without gravity's assistance meant, in effect, performing daily sit-ups—resulting in abdominal muscles that grew stronger in space than they had ever been on Earth.) Today's spacefarers follow a rigorous, rather tedious exercise regimen to ensure that the major muscles in their arms and legs will be fit for full gravity. Soviet crews work out for at least two hours a day, pedaling a stationary bicycle, pounding a treadmill, and building up their arm muscles with elastic cords. Aboard Skylab, the Americans increased their workouts from half an hour per day on the first mission to ninety minutes a day on the final stint.

Less well understood than muscular atrophy is the condition called disuse osteoporosis. In space, scientists have found, the bones themselves deteriorate, losing vital calcium and thus the strength that protects them from fractures. Cosmonauts domiciled on *Salyut 6* for 175 days, for example, lost about eight percent of the calcium in their bones. Researchers are not certain whether this demineralization levels off after a certain period, nor do they know whether weakened bones fully recover once a space traveler returns to Earth. Crews on future space stations will have the task of studying long-term effects in greater detail and finding the best ways to combat them.

LIFE SUPPORT

Orbiting stations are vital to the development of another key facet of space exploration: self-sustaining life-support systems. NASA calculates that a person consumes 2.2 pounds of oxygen, 1.3 pounds of food, and nearly 6 pounds of water each day. A three-person crew on a year-long mission would require nearly five tons of food, water, and fresh air. Crews aboard space stations in low Earth orbit can count on receiving these supplies from Earth, although the cost of sending up cargo ships is steep. For journeys to distant points, however, spacefarers will need what is called a closed-loop system, capable of producing food, oxygen, and water as well as recycling wastes.

Station engineers have already made some progress. Since *Salyut 6*, for example, Soviet stations have been equipped with a water reclamation system that extracts moisture from the cosmonauts' exhalations and purifies it for drinking. The contrivance reclaims about 50 percent of the water on board, thus reducing the weight of stored materials from 10.7 tons to 2 tons. On Mir, cabin air is also recycled to produce fresh oxygen, and a system known as Electron uses chemical electrolysis to decompose water into its constituent elements, oxygen and hydrogen.

Looking ahead, scientists agree that they must achieve a system in which water, oxygen, and food could be produced and recycled in a completely self-sustaining cycle. Such a procedure would depend, in space as on Earth, upon plant life. Unfortunately, plants have so far proved less amenable to life in space than the humans who have carried them there. On the early Soviet missions, crews sowed a variety of crops, such as onions, tomatoes, and wheat, only to see some of them wither and others drown in their own weightless fluids. Some sprouted but refused to grow seeds. *Salyut 7* improved the situation with additional lighting and a centrifuge to supply the plants with artificial gravity. Finally, during a 1982 mission aboard *Salyut 7,* the cosmonauts brought one species—a quarry weed, arabidopsis—through its entire life cycle inside a special device called a Phyton. The container provided special nutrients and artificial light and filtered out impurities in the air. When cosmonaut Svetlana Savitskaya arrived on her assignment three months into the mission, her colleagues were able to present her with a bouquet of arabidopsis flowers.

Encouraged by these Soviet successes, the United States began to move ahead with its own plans for an advanced space station. After Skylab, as the country recovered from the costs of a war in Vietnam and weathered a recession, national priorities had turned toward the development of an economical, reusable spaceplane—a shuttle that could put commercial satellites into orbit as well as serve scientific ends. Scientists interested in a space station also viewed the shuttle as a prototype for a fleet of ferries to take people, supplies, and building materials back and forth during station construction. On April 12, 1981, the United States launched the first such orbital transport, the space shuttle *Columbia.*

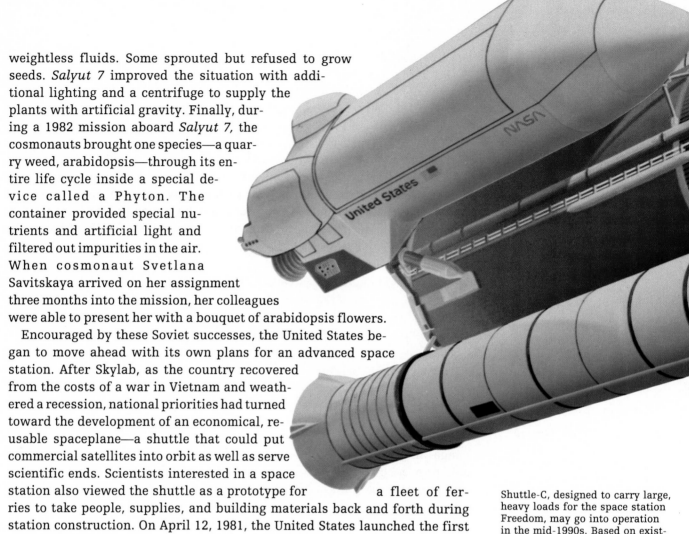

Shuttle-C, designed to carry large, heavy loads for the space station Freedom, may go into operation in the mid-1990s. Based on existing shuttle technology, shuttle-C would be a cargo-only vehicle capable of lifting forty to seventy-five tons into space.

A FERRY TO SPACE

Columbia and its successors are hybrids, blending airplane and rocket in a bulky package. Strapped to each winged orbiter are two rocket boosters and a huge fuel tank. During the journey into space, the boosters and tank fall away, and the orbiter coasts around Earth at 17,500 miles per hour. When the journey is over, the orbiter's engines engage once more, performing a "deorbit burn" that slows the craft and aims it into the atmosphere. There, amid the thick currents of air, the spaceship decelerates further, until it descends at last to Earth and glides to a halt on a runway. After some refurbishing, the shuttle can be taken on another mission.

For five years and twenty-four flights, shuttles lifted off from Florida's Cape Canaveral virtually without incident. But the disastrous explosion of the shuttle *Challenger* on a frigid day in January 1986 forced NASA into a long period of soul-searching. Two years later the space agency emerged from this introspection with a renewed commitment to interplanetary exploration—an effort that will initially focus on a space station. Joining NASA in its efforts on the cusp of the twenty-first century will be an international assembly of space ferries. Britain, Germany, the European Space Agency, and Japan have all drawn up plans for sophisticated transport vehicles *(pages 29-33)*. Engineers in the United States have also proposed following up on passenger shuttles with vehicles designed specifically to ferry heavy cargoes. One shuttle derivative, dubbed the shuttle-C, would be an unpiloted carrier launched by rocket engines powerful enough to boost payloads triple the weight of the present shuttles' capacity. NASA and the U.S. Department of Defense are also studying concepts for a family of heavy-lift vehicles known as the Advanced Launch System, which might put payloads into orbit fairly cheaply. In the meantime, the Soviet Union came out in 1988 with its own shuttle *Buran,* borne on the massive booster rocket Energia.

VISION FOR THE 1990s

Space stations now on the drawing boards will pick up where Salyut, Mir, and Skylab left off. For the United States, a major goal is to establish a permanent staffed base in orbit in the 1990s. Such a base, said President Ronald Reagan in 1984, would yield "quantum leaps" in scientific research. Four years later, Reagan gave the station a name—Freedom—and NASA established a space station program office in Reston, Virginia, to oversee the project.

From conception to operation, Freedom presents a daunting logistical challenge. Simply to construct the parts on Earth will take six years; three more will be needed to launch and assemble them in orbit. NASA estimates that the first phase of the station will be complete in the late 1990s. The cost, including development, launch, assembly, and operations, may run to $30 billion; 50,000 people across the country will be involved at the peak of the process. Once aloft, Freedom is expected to function for three decades, circling Earth every ninety minutes at an altitude of 220 miles. It will be staffed by crews of eight, who will eventually stay aboard for up to six months at a time.

Structurally, the space station will be simple but flexible; its modular design will allow parts to be added or removed as necessary. At the core of the finished station will be a cluster of pressurized modules attached to a 500-foot-long latticework transverse beam. The crew will live and work in the two largest modules, each about forty-five feet long and fifteen feet in diameter, which will be supplied by the United States. Japan and the multinational European Space Agency will furnish several smaller laboratory units, and Canada plans to contribute a mobile robotic arm that will assist in assembly and repairs. At either end of the long beam will be giant solar panels, generating a total of seventy-five kilowatts to run the station's systems and experiments. In the long run, NASA hopes to add on several free-flying platforms, near the main station but unaffected by its vibrations, which would provide suitable areas for delicate experiments and observations.

The space agency's goals are broad. By 1998, after two years of operating with pioneer crews of four, the station is scheduled to be working with a full complement of staff. Once on board, the space dwellers will conduct experiments into the effects of microgravity on a variety of materials, as well as on living creatures. From their favorable vantage point above the atmosphere, they will also turn telescopes on both the Earth and the stars. By the twenty-first century, the base should be a thriving enterprise.

A SETTLEMENT IN ORBIT

The crews who spend half a year aboard Freedom will live in ingeniously designed, if not roomy, quarters. Inside the gleaming habitation module, about the size of a house trailer, each resident will have a private compartment containing a sleeping sack mounted on the wall, a television, a stereo, a computer terminal, and telephone and video links to Earth. Elsewhere in the living module will be a bathroom, a shower, a gym, a combination dining and social area, and a clothes washer and dryer.

Meals should be a vast improvement over the food available on the first space stations. The international cuisine, ranging from tortillas to beef stroganoff, will be highly seasoned to adjust for the diminished sense of taste that seems to accompany the redistribution of the body's fluids in microgravity. Crew members will be able to design their own menus by ordering from a voice-activated computer, which will keep track of the galley's inventory, and they will eat at a table near a large window overlooking the Earth below.

During the workday, Freedom's residents will be busy in the adjoining laboratory module. At any time, more than eighty projects and experiments may be lined up in racks along the walls and on the floor. Some of the work will involve materials processing. On Earth, substances with different densities and temperatures separate under the influence of gravity, requiring complex and expensive procedures to combine them. The microgravity of a space platform will allow engineers to try unique methods of manipulating these items, forging extra-strong metal alloys, purifying medical products, and growing large, uniform crystals that could become invaluable to the electronics industry back home.

Other projects in the laboratory module will focus on life sciences. To study the role of gravity in the development of plants and animals, the Freedom crew will raise various kinds of cells, tissues, plants, and even small animals, such as rats, in a special centrifuge that simulates a range of gravities. The hope is that the experiments will help scientists determine how to use artificial gravity to offset the potentially destructive effects of weightlessness on both food crops and human travelers. Also in the interest of future space travel, the Freedom settlers might raise an alga known as chlorella. Unlike more complex plants, which struggle for survival in space, chlorella actually grows more rapidly in orbit than on Earth. The space-hardy organism, or another alga like it, could serve as the basis of a closed-loop life-support system, manufacturing much of the oxygen, water, and protein necessary to sustain spacefarers journeying to distant destinations. Unfortunately, scientists have

An International Space Fleet

Aeronautical engineers have long dreamed of a spacecraft that could take off and land horizontally, using a conventional airplane runway, yet be powerful enough to achieve mach 25 (twenty-five times the speed of sound), the speed required to break free of Earth's gravity and go into orbit. Such a vehicle—nicknamed an orbital express—offers two advantages over a rocket-launched orbiter: It is fully reusable (no throwaway booster rockets), and it does not need elaborate launch facilities.

This craft may one day be a reality. As shown here and on the following pages, plans for orbital expresses are on the drawing boards in several countries. In the near term are systems that, like the American space shuttle, launch the orbiter on expendable rocket boosters. Most of them will carry crews, numbering from three to ten members, although the systems are also designed to launch unpiloted cargo carriers. The crews might stay in orbit for several days to conduct experiments or make shorter trips to deliver building materials, supplies, or scientific equipment to a space station or other orbiting craft. The orbital expresses would also be used for launching satellites and picking up astronauts, experiments, or other spacecraft for return to Earth.

The Soviets have already flown the first of these interim spaceplanes: the shuttle *Buran* ("snowstorm"), which closely resembles the American shuttle in size and configuration. The multinational European Space Agency and the Japanese are designing minishuttles of markedly smaller size. Called *Hermes* and HOPE, respectively, the orbiters would rendezvous with international additions to the American-built space station, Freedom.

Looking ahead to the more distant future, West Germany and the United States are studying orbital expresses that would use new types of jet engines. The German spacecraft is the two-stage *Sänger*. More technically advanced is the American design, the National Aero Space Plane project's single-stage-to-orbit X-30. For both the *Sänger* and the X-30, the goal is to turn flight to orbit into an extension of aviation—making it as routine, and as inexpensive, as flight in high-speed aircraft is today.

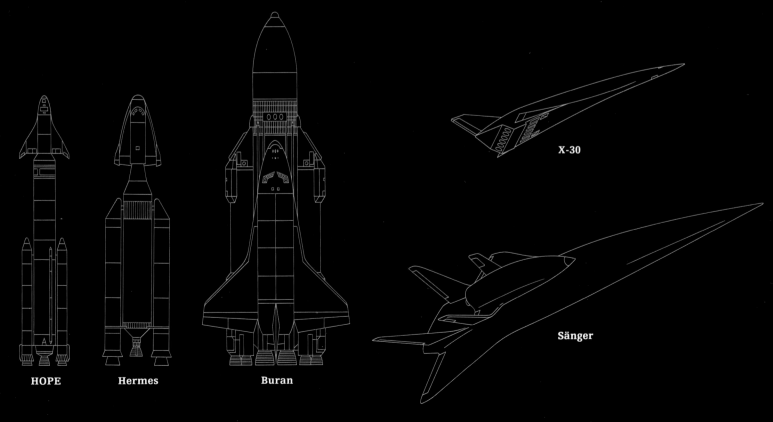

HOPE   Hermes   Buran   X-30   Sänger

ROCKET LAUNCHES, RUNWAY LANDINGS

The three orbiters shown here, all scheduled to be operational by 1998, bear a family resemblance to one another and to the American space shuttle. Each rides a first stage of expendable rockets, and each is designed to make its return landing on a runway.

The piloted Soviet shuttle *Buran (below)* is boosted into orbit aboard the heavy-lift Energia rocket, a powerhouse with 8.9 million pounds of total thrust. De-

Buran rides into orbit aboard the powerful Energia rocket—four liquid oxygen and hydrogen main engines and four oxygen and kerosene strap-on boosters. Placing the main engines on the expendable booster saves weight on the orbiter for the return payload. The delta-winged *Buran* is equipped with a pair of powerful jet engines that let it "go around"—to make another landing attempt if it misses on its first try.

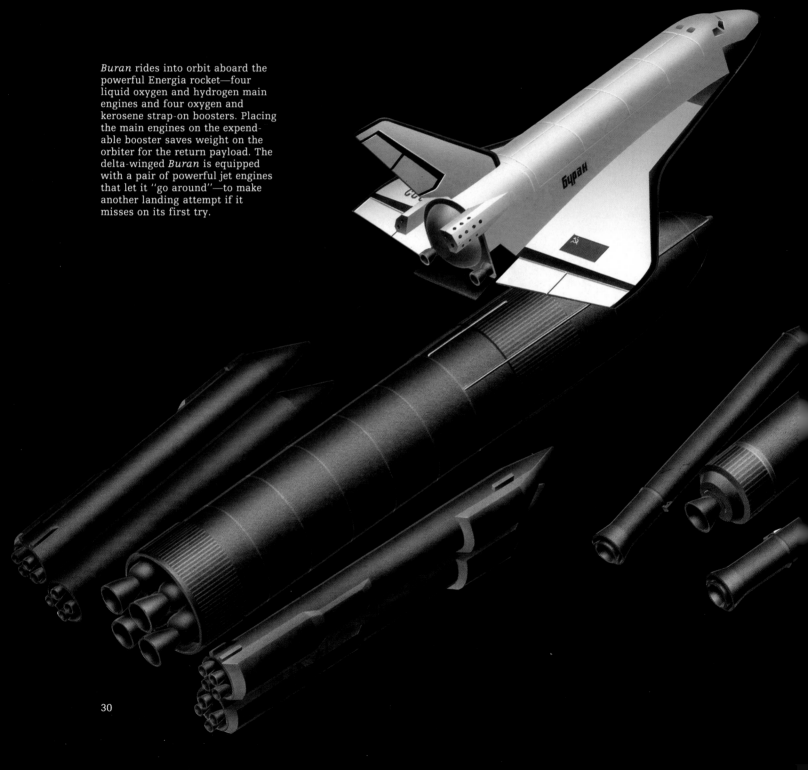

signed to carry up to ten crew members, *Buran* can haul thirty tons of cargo into low Earth orbit and return twenty tons of material to Earth.

The European Space Agency's system relies on Ariane 5, a heavy-lift rocket that is one of a series of satellite launchers. Ariane 5 will boost *Hermes*, which carries a crew of three and up to three tons of payload. *Hermes* will be able to dock with Columbus, Europe's free-flying laboratory, and with the international space station Freedom.

Japan's system features the H-2 launch vehicle and the unpiloted minishuttle HOPE (H-2 Orbiting Plane), which will be able to carry 1.2 tons in the cargo bay. HOPE is intended to supply not only the Japanese experiment module on Freedom but orbiting scientific platforms as well.

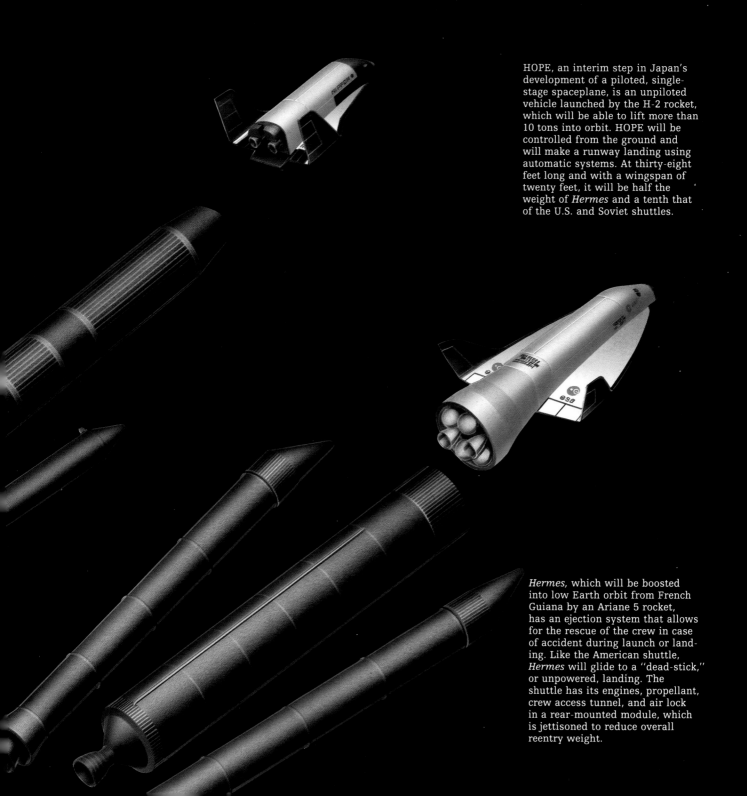

HOPE, an interim step in Japan's development of a piloted, single-stage spaceplane, is an unpiloted vehicle launched by the H-2 rocket, which will be able to lift more than 10 tons into orbit. HOPE will be controlled from the ground and will make a runway landing using automatic systems. At thirty-eight feet long and with a wingspan of twenty feet, it will be half the weight of *Hermes* and a tenth that of the U.S. and Soviet shuttles.

Hermes, which will be boosted into low Earth orbit from French Guiana by an Ariane 5 rocket, has an ejection system that allows for the rescue of the crew in case of accident during launch or landing. Like the American shuttle, *Hermes* will glide to a "dead-stick," or unpowered, landing. The shuttle has its engines, propellant, crew access tunnel, and air lock in a rear-mounted module, which is jettisoned to reduce overall reentry weight.

Spaceplanes for Flying into Orbit

To do away with the rockets that prevent a runway takeoff, a spaceplane needs an air-breathing engine that can power it to just below escape velocity, or mach 25. However, the fastest jet in the air today—the SR-71, known as the Blackbird—flies just a little faster than 2,000 miles per hour, or only mach 3; above that speed, the turbines that drive the compressors needed to force air into the craft's turbojet engines overheat.

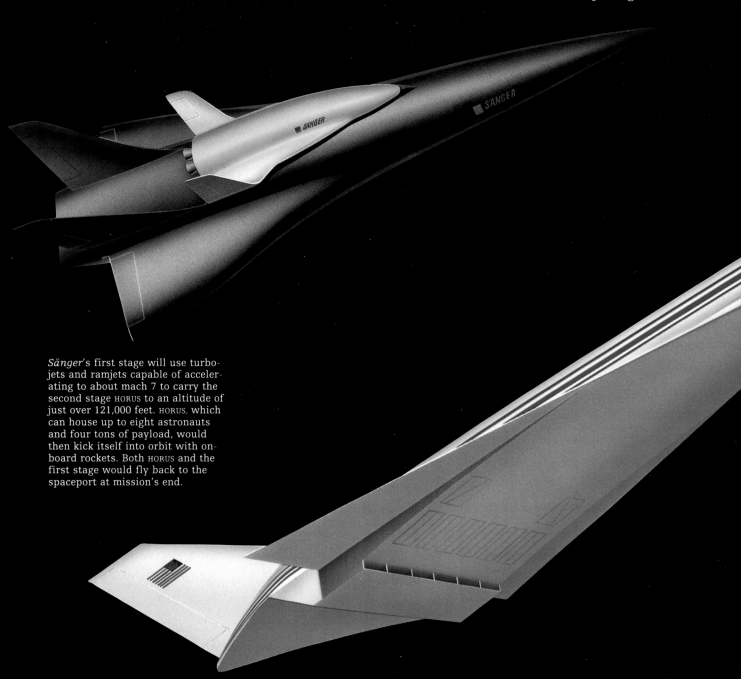

Sänger's first stage will use turbojets and ramjets capable of accelerating to about mach 7 to carry the second stage HORUS to an altitude of just over 121,000 feet. HORUS, which can house up to eight astronauts and four tons of payload, would then kick itself into orbit with onboard rockets. Both HORUS and the first stage would fly back to the spaceport at mission's end.

The two designs for orbital expresses shown here—West Germany's two-stage *Sänger* and NASA's single-stage X-30—attack the problem with a combination of turbojets, more advanced engines, and small rockets. The *Sänger*'s first stage would take off from a runway in Europe and cruise over the Atlantic before accelerating to about mach 7. It would then launch a rocket-powered second stage: either the piloted HORUS (hy-personic orbital upper stage) or the unpiloted CARGO (cargo upper stage), which can carry fifteen tons of payload. The X-30 would rely on a new kind of engine—the scramjet, for supersonic combustion ramjet. Scramjets would require the airframe and engine of the spaceplane to be fully integrated in order to reduce drag while making the most efficient use of the supersonic air flowing over the aircraft.

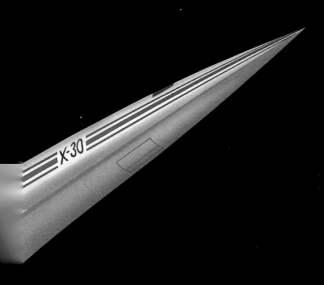

The entire underside of the X-30 serves as part of its scramjet propulsion system. Forward of the boxy scramjet modules, the undersurface helps compress incoming air. Aft of the scramjets, the underbody curves sharply upward, allowing the exhaust to expand and produce thrust. Small rocket engines *(not shown)* may give the X-30 a final boost to mach 25.

A Trio of Engines

A turbojet's rotating compressor pulls in and compresses air, and exhaust gases spin a turbine to power the compressor. But above mach 3, the turbine overheats and the engine loses power. A ramjet has no compressor or turbine but needs an initial turbojet boost to mach 3 so that the engine can compress air by simply ramming into it. The resulting shock wave slows the incoming supersonic air, allowing fuel to burn subsonically, a process that works well up to mach 6. Still in the developmental stage, a scramjet is a ramjet designed to handle supersonic air. In theory, incoming air compresses without overheating, all the way to orbital velocity—mach 25.

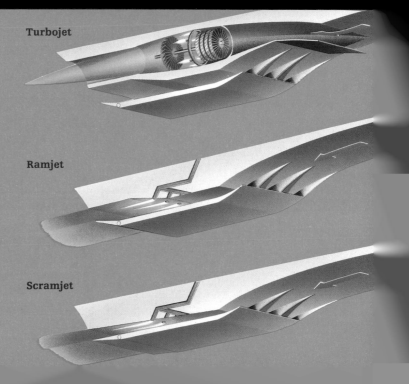

Turbojet

Ramjet

Scramjet

not yet discovered a way around a major obstacle: chlorella is notoriously difficult for humans to digest.

Early in the twenty-first century, a second, free-flying space platform will be built. Isolated from Freedom to protect the main station from vibrations, this will be the in-space construction facility, a space hangar in which ships to the Moon, Mars, or even more distant corners of the Solar System will be built. With the completion of this hangar, the space station will become the home port of the first interplanetary travelers.

COLONIZING

Sophisticated as today's technology is becoming, these stations and vehicles are still far removed from the sprawling space utopias envisioned by early enthusiasts. Some modern thinkers about space travel feel that stations such as Freedom do not go nearly far enough in exploiting the possibilities of life in space, especially as humans deplete Earth's limited resources. The pioneer and chief philosopher in this school of thought is American physicist Gerard O'Neill, now professor emeritus at Princeton University.

In the 1960s, O'Neill had been a finalist in NASA's short-lived scientist-astronaut program. By the end of the decade he was convinced that orbiting habitats could support life better than Earth itself. Ultimately he wrote a book called *The High Frontier,* published in 1976, which spelled out plans for constructing space colonies on a grand scale. The colonies would be balanced in space at a so-called libration point—one of five locations between Earth and its moon where the gravitational pull of the two bodies is balanced. Any object at a libration point should maintain a stable position in relation to Earth and the Moon *(page 57).* The colonies would use materials transported from both Earth and the Moon. Island One, as O'Neill called the first colony, might be a sphere nearly a mile in diameter at its equator and inhabited by 10,000 colonists. By the time Island Three is built, he theorized, colonies could be half as large as Switzerland, with populations in the millions.

According to O'Neill, a space colony such as Island One would create artificial gravity by rotating once every thirty-two seconds. Rising around the equator would be apartments, parks, and shops. Other services and industries might be located "underground" (the colony would be layered with soil) or in a central, low-gravity sphere. Day-to-day life would be similar to that on Earth in a climate like Hawaii's. O'Neill envisioned a very practical purpose for his islands in space: There, residents would build and maintain solar satellites to harness the Sun's energy and beam it back to Earth.

Although NASA recognized O'Neill's work as path breaking, his colonies remain no more than dreams. He is not the only dreamer. Other visionaries include members of the National Space Society, an American organization that, until its merger with the National Space Institute in 1987, was known as the L-5 Society, taking its name from one of the five libration points. The National Space Society is planning for a time when humans will establish a space colony at the fifth libration point, one of the most stable. Others carry

COMPONENTS OF A LIGHT, STRONG SPINE

Node

Joint Mechanism

Strut

The backbone of most structures assembled in space, from small satellites to massive space stations, is a strong but simple framework called a truss. Used for hundreds of years to bear terrestrial loads such as bridges and roofs, trusses owe their strength and stiffness to interlocked triangles that distribute stress across the entire frame.

For the American-built space station Freedom, prefabricated modules for living, working, service, and storage are designed to be attached to a square truss, 508 feet long and 16 feet on a side. The truss incorporates just three kinds of parts *(above):* long, two-inch-diameter tubes, or struts; grapefruit-size aluminum balls called nodes, each pocked with twenty-six threaded holes; and five-inch-long joint mechanisms used to connect struts and nodes.

The parts are easy to manipulate even with bulky spacesuit gloves, as demonstrated on a 1985 space shuttle flight when crew members put together a small-scale experimental truss. The astronauts screw a joint mechanism into a node, then couple a strut with the other end of the joint mechanism. One half-twist secures the strut temporarily; a second half-twist locks it in place.

To save weight, the struts are made of graphite fibers embedded in epoxy resin, a composite twice as stiff as aluminum but only half as heavy. The strut walls, one-sixteenth of an inch thick, are coated inside and out with a paper-thin aluminum skin to prevent warping caused by the Sun's heat. The skin also protects the struts from erosion by the rarefied orbital atmosphere.

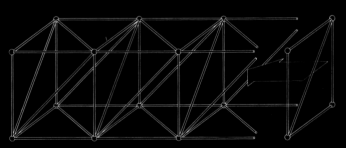

The struts that make up the latticework of Freedom's truss are assembled into diagonally braced squares *(left),* which form the sides of open cubic frames, or bays, locked end to end. The truss grows a piece at a time, with lateral and diagonal struts added in a carefully prescribed sequence.

the dream even further, imagining the day when huge traveling colonies will rocket toward unknown solar systems to seek communities of other living creatures. No one can say for sure whether this will happen, or even if humans will develop the technology to make such journeys possible. But it is worthwhile remembering how much of modern space science draws on ideas spawned by early visionaries: von Braun's winged spacecraft, for example, and Clarke's space stations assembled in orbit. It seems certain, at least, that in the next few decades more people will travel into space than ever before, to live and work on Freedom and other stations of its kind. From these bases, beyond the pull of gravity, the possibilities can only expand.

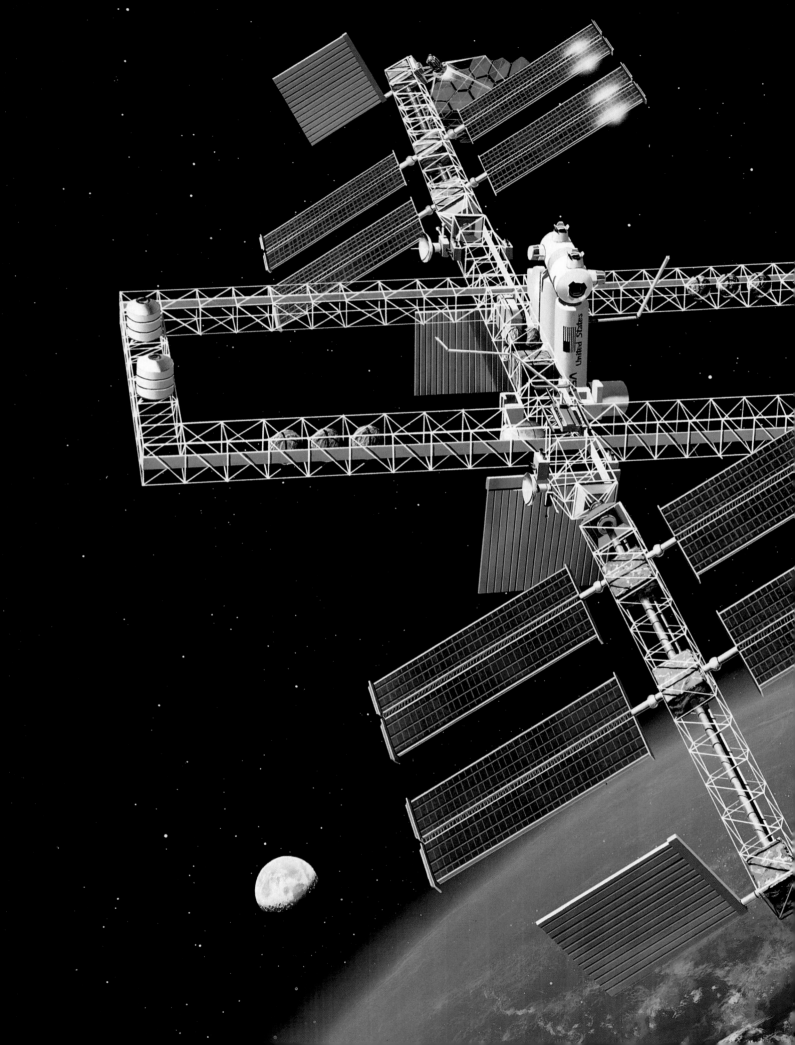

G leaming in unfiltered light, its winglike solar panels and parabolic solar mirrors poised to follow the Sun, the American space station Freedom is depicted here as it will look in Earth orbit shortly after completion of its second phase of construction early in the twenty-first century. From Earth, 250 miles below, the station will be visible as a fast-moving object in the night sky, apparently larger than Venus and brighter than the Moon.

The building of Freedom—traced in part on the following pages—is scheduled to begin in the mid-1990s with more than a dozen space shuttle missions that will haul components into orbit. Within a few months, engineers and scientists will begin to arrive, at first staying for only a couple of weeks; by the end of the second year, assignments will last up to six months. When the first experiments in the station's international laboratory complex get under way, researchers from the United States, Canada, Europe, and Japan will study Earth and the heavens from an extraordinary vantage point. They will also perform experiments unique to space, such as growing perfect crystals.

Freedom is meant to be a gateway to interplanetary space and beyond—a refueling and repair station for satellites and transfer vehicles, as well as a place where scientists can study the effects on humans of extended stays in space. Over the course of its working life of twenty to thirty years, the station may be reconfigured for additional uses. The living and working areas, for example, will be equipped with connecting nodes that readily accommodate expansion; extra attachment points on the keels could link to hangars for servicing spaceships. Station astronauts, working at a co-orbiting construction facility *(pages 45-47)*, will build huge ships for voyages to the Moon, Mars, and other planets—prelude to a future when a similar station could launch vessels to the stars.

Supplying a Celestial Construction Site

Like any big construction project, Freedom will require a steady flow of building supplies. The delivery vehicles will be space shuttles, arriving every other month with as much as 40,000 pounds of station components. To minimize the delivery cost, logistics experts will use computers to assess every ounce and cubic inch of the objects in a given payload and optimize their positioning in the sixty-by-fifteen-foot cargo bay. Doing this will also maintain the shuttle's center of gravity in case it must make an emergency landing shortly after launch. Yet another benefit of the analysis will be to ensure that, once the shuttle is in orbit and construction of the station has begun, pieces will come to hand in logical order.

The first shuttle flight will carry an impressive array of components. For example, each of four truss boxes will hold thirteen struts and several nodes *(page 35)*, the basic building blocks for the station's latticework structure. A pair of solar photovoltaic panels—the first of four such sets—will generate enough electricity to run computer-controlled guidance and communications systems for the rudimentary station. Thrusters in the reaction control system will allow the station crew and ground control to maintain the station's attitude and adjust its orbital altitude. To conserve shuttle fuel, for instance, rendezvous with the station is timed to occur when atmospheric drag has lowered perigee, or the lowest point of the station's orbit, to an optimal altitude. Once the shuttle departs, the station can use its thrusters to kick into a higher orbit to compensate for atmospheric drag.

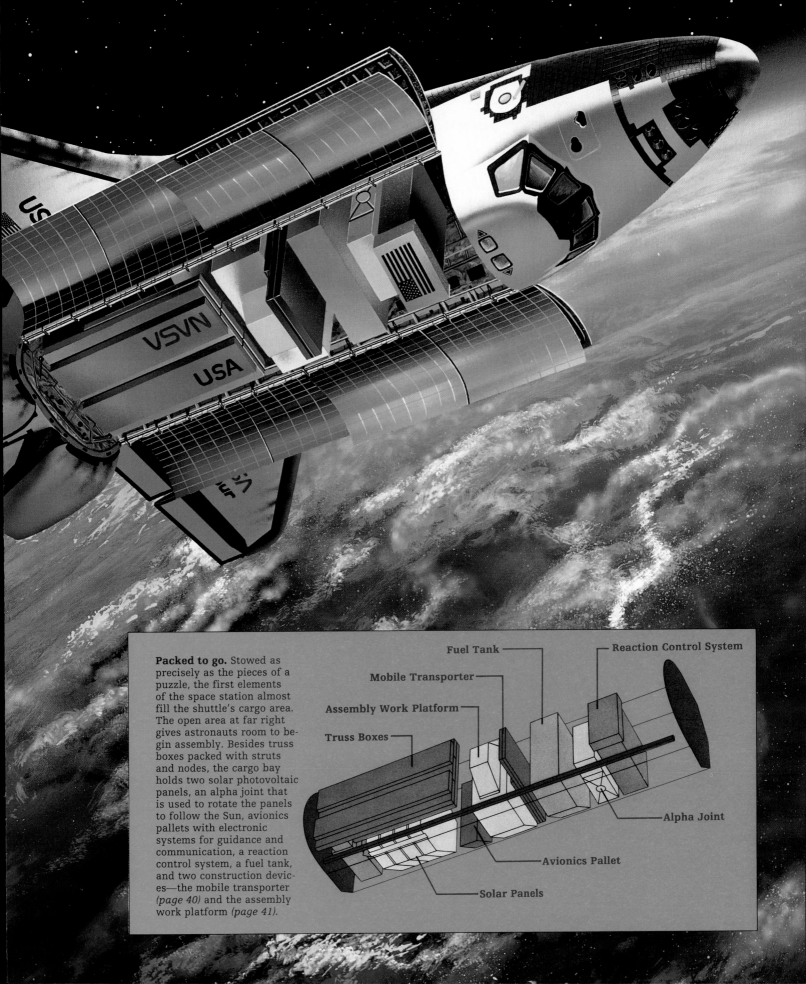

Packed to go. Stowed as precisely as the pieces of a puzzle, the first elements of the space station almost fill the shuttle's cargo area. The open area at far right gives astronauts room to begin assembly. Besides truss boxes packed with struts and nodes, the cargo bay holds two solar photovoltaic panels, an alpha joint that is used to rotate the panels to follow the Sun, avionics pallets with electronic systems for guidance and communication, a reaction control system, a fuel tank, and two construction devices—the mobile transporter *(page 40)* and the assembly work platform *(page 41).*

Fuel Tank

Reaction Control System

Mobile Transporter

Assembly Work Platform

Truss Boxes

Alpha Joint

Avionics Pallet

Solar Panels

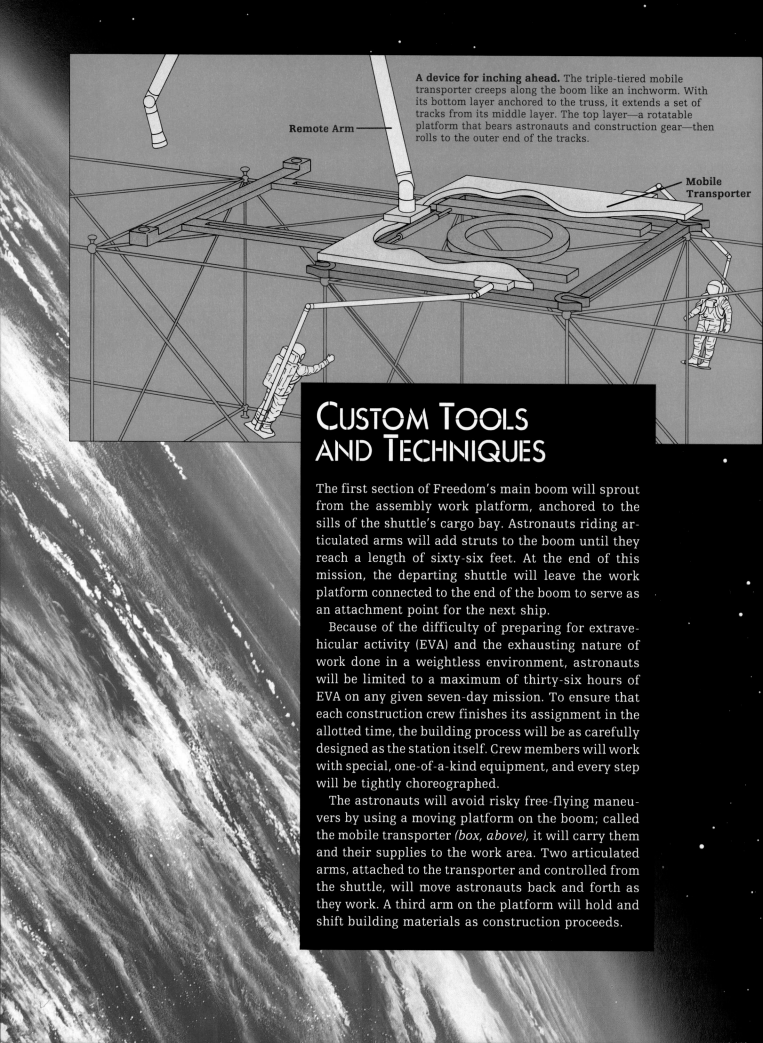

A device for inching ahead. The triple-tiered mobile transporter creeps along the boom like an inchworm. With its bottom layer anchored to the truss, it extends a set of tracks from its middle layer. The top layer—a rotatable platform that bears astronauts and construction gear—then rolls to the outer end of the tracks.

Remote Arm

Mobile Transporter

CUSTOM TOOLS AND TECHNIQUES

The first section of Freedom's main boom will sprout from the assembly work platform, anchored to the sills of the shuttle's cargo bay. Astronauts riding articulated arms will add struts to the boom until they reach a length of sixty-six feet. At the end of this mission, the departing shuttle will leave the work platform connected to the end of the boom to serve as an attachment point for the next ship.

Because of the difficulty of preparing for extravehicular activity (EVA) and the exhausting nature of work done in a weightless environment, astronauts will be limited to a maximum of thirty-six hours of EVA on any given seven-day mission. To ensure that each construction crew finishes its assignment in the allotted time, the building process will be as carefully designed as the station itself. Crew members will work with special, one-of-a-kind equipment, and every step will be tightly choreographed.

The astronauts will avoid risky free-flying maneuvers by using a moving platform on the boom; called the mobile transporter *(box, above),* it will carry them and their supplies to the work area. Two articulated arms, attached to the transporter and controlled from the shuttle, will move astronauts back and forth as they work. A third arm on the platform will hold and shift building materials as construction proceeds.

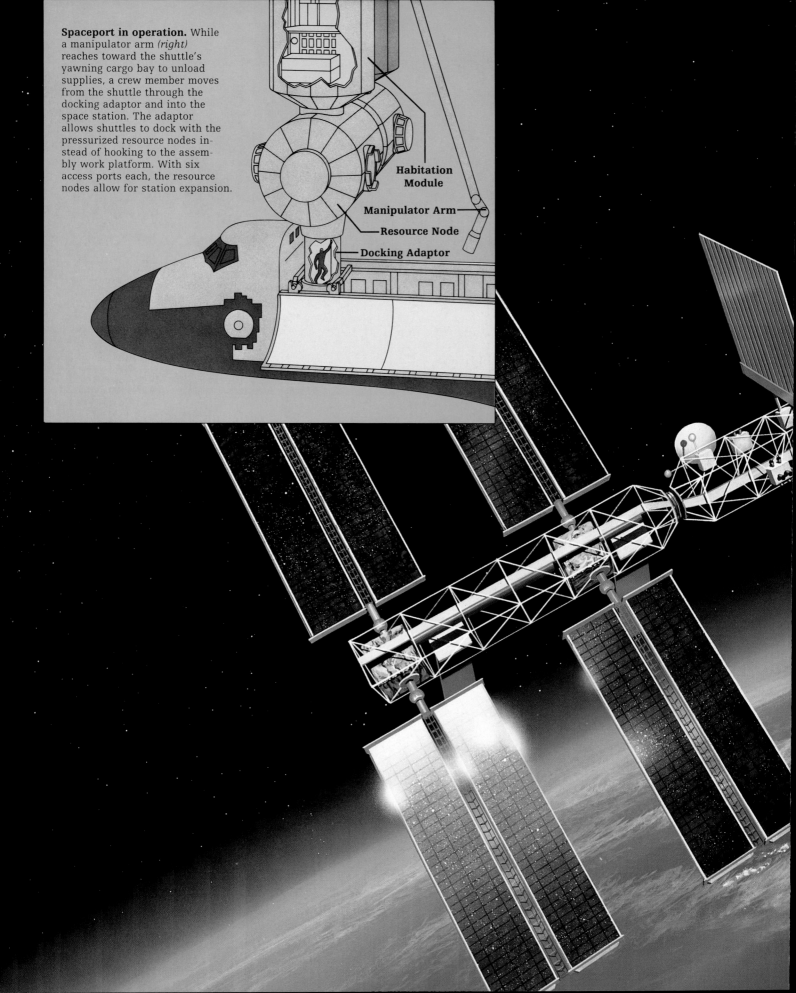

Spaceport in operation. While a manipulator arm *(right)* reaches toward the shuttle's yawning cargo bay to unload supplies, a crew member moves from the shuttle through the docking adaptor and into the space station. The adaptor allows shuttles to dock with the pressurized resource nodes instead of hooking to the assembly work platform. With six access ports each, the resource nodes allow for station expansion.

Habitation Module

Manipulator Arm

Resource Node

Docking Adaptor

WRAPPING UP PHASE ONE

As the first phase of construction draws to a close, the space shuttle on the final outfitting mission will still be docked with Freedom, connected to a so-called resource node that in turn is linked to the habitation module, where the eight persons in the station crew eat, sleep, and relax off duty. More resource nodes will connect the living quarters with three other modules, each about the size of a mobile home, that provide working space for scientific experiments. The United States laboratory will be positioned next to the habitation module; across the main boom will lie the European Space Agency's scientific module and the Japanese experimental module (JEM). The JEM will have its own manipulator arm and external platform, a "back porch" where objects can be tested for the effects of space exposure.

The eight solar photovoltaic wings at the ends of the boom will generate 75,000 watts—enough to feed twenty-five all-electric homes back on Earth. During sunlight hours, the assembly will not only supply the station but also charge batteries that will be used when the station is on the dark side of the Earth. Six radiator panels attached to the boom will dissipate heat from the solar panels and the habitable spaces.

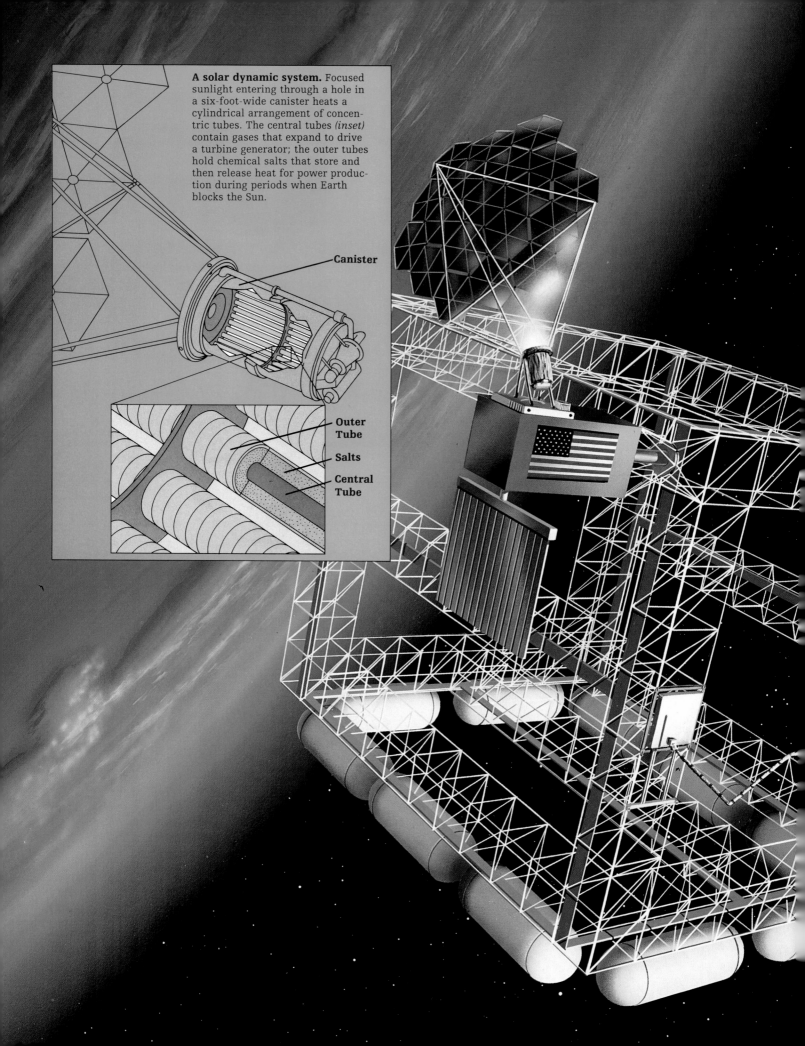

A solar dynamic system. Focused sunlight entering through a hole in a six-foot-wide canister heats a cylindrical arrangement of concentric tubes. The central tubes *(inset)* contain gases that expand to drive a turbine generator; the outer tubes hold chemical salts that store and then release heat for power production during periods when Earth blocks the Sun.

Canister

Outer Tube

Salts

Central Tube

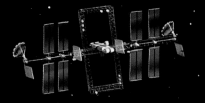

BUILDING AN ANNEX BY REMOTE CONTROL

Traveling in the same orbit but trailing thirty miles behind the Freedom space station, a companion structure called the in-space construction facility, shown here as it nears completion, will serve as a factory for building mammoth spacecraft. Confining the job to this venue will protect Freedom's sensitive laboratory experiments and delicate scientific equipment from the bumps and vibrations of shipbuilding.

Although workers will commute from Freedom to the factory when spaceship assembly gets under way, the construction of the co-orbiting facility itself will be done largely by remote control, minimizing potentially hazardous extravehicular activities. As illustrated at left, a robotic mobile transporter (RMT) could assemble a truss, guided by an operator aboard Freedom who monitors the work through a video camera on the RMT platform. Pods attached to the trusses would hold supplies.

Parachute-like mirrors at the corners of the facility will focus sunlight into power-generating solar dynamic systems *(box, far left)*. In addition to supplying nearly three times as much power as a photovoltaic wing, the solar dynamic systems will eliminate the need for periodic battery replacement.

ON THE JOB IN AN ORBITING SHIPYARD

In the scene at right, work in the completed construction facility progresses on an aerobrake that will protect a lunar spacecraft when it enters Earth's atmosphere at the end of a mission. The ship itself is tethered within the co-orbiting facility's cavernous bay, shielded from the intense sunshine of space by sheets of gold foil mounted on the side trusses.

Astronauts inside the facility guide a computer-controlled crane, an articulated truss 311 feet long. The crane's joints turn on special offset hinges that allow a 180-degree range of motion, giving the huge tool access to any location in or around the facility. With its grasping end—a duplicate of the remote arm of the mobile transporter originally used to build the crane itself—the crane can maneuver a hexagonal piece of the aerobrake's heat shield into place. After the pieces are fixed to nodes on the supporting structure, astronauts will likely use a telerobotic arm to weld them in place. Upon completion, the spacecraft will be transferred to the Freedom space station for fuel and provisions before undertaking its round-trip flight to the Moon.

The 6,500-foot-high North Massif looms over the rocky plains of the Sea of Serenity, dwarfing the figure of *Apollo 17* astronaut Harrison Schmitt in a photograph taken during the last piloted mission to the Moon.

he space station has receded to a brilliant speck, soon lost in the blue-white glow of Earth. In the lunar transport's window, the Moon draws near, large and silent, its gray-brown face blasted and pocked like a huge battlefield. The transport enters lunar orbit, and after two hours, traveling from sunlight to darkness and back to sun again, the four passengers spot a few signs of habitation: light glinting off metal in the Mare Tranquillitatis, or Sea of Tranquillity. Donning bulky spacesuits, the men and women clamber through an air lock into the automated lunar lander, which separates from the transport, fires its rockets, and starts the seventy-mile descent. Gravity's clumsy hand grabs the lander's occupants as the small spacecraft skims over Tranquillity and its landscape of harsh light and deep shadows.

Ahead, the Moon base heaves into view. To the left of an open-pit mine are the igloolike living quarters, each with its own solar panels. Beyond lies the nuclear generator that is the base's main power supply.

A trail scratched on the surface leads over the horizon from the igloo cluster to a circular landing area. As the craft eases down, its retrorockets kicking up a shroud of dust, the new arrivals catch a brief glimpse of the lunar rover, approaching at the head of its own dust cloud. The driver, a member of the construction crew, is doubtless grateful that his months-long shift is almost over and looks forward to his own departure a few days hence. But as the visitors prepare to step down onto the lunar surface, they are contemplating the past, not the future. Only a few miles from where they have landed, near the southwest corner of Tranquillity, are the sharp-edged, corrugated footprints of Edwin "Buzz" Aldrin and Neil Armstrong. Having lasted unweathered for thirty-eight years, they could well remain unchanged for thirty-eight thousand more. . . .

. . . The year of this imagined scene is 2007—about a decade after the United States, in cooperation with several other nations, will begin its return to the Moon by mapping suitable lunar base sites with robot spacecraft, and less than a decade before the next planned leap in space: a manned mission to Mars. The lunar base will provide NASA with the knowledge, and possibly some of the materials, it will need to make that leap.

For instance, one of the scientific crew's most important tasks will be to operate a biomedical laboratory to study the effects of living and working for sustained periods on an alien surface. The lunar outpost will also be a testing ground for the kinds of spacecraft, surface vehicles, and engineering technology that might be used on the Mars mission. For example, oxygen extracted from the lunar soil, in addition to providing air for breathing, will be processed into liquid rocket propellant. At first, it will supply transports to and from lunar orbit. Later, it might power ships for the nearly three-year round-trip mission to Mars.

IN THE REALM OF IMAGINATION

When the twentieth century was young, the idea of occupying the Moon was the province of visionary rocket engineers and science-fiction writers, who believed that Earth's satellite could provide valuable resources for both scientists and explorers. As early as 1920, American rocket pioneer Robert Goddard suggested mining extraterrestrial sources, such as the Moon, for fuel and other materials needed for space travel. In 1950, writer Arthur C. Clarke published an article expanding on this concept. Clarke proposed that the mined substances be shot from the lunar surface by means of an electromagnetic accelerator.

Until the dawn of the space age, however, neither engineers nor fiction writers had much beyond their own imaginations on which to base their beliefs. But with the firing of a Russian rocket probe in October of 1959 came a remarkable advance in lunar geology. The probe sent back the first pictures of the Moon's far side, showing scientists a rougher surface with markedly fewer plains than that of the near side. Over the next several years, as NASA geared up for the Apollo landings, American probes took thousands of pictures of the lunar landscape, photographing virtually all of the Moon's face. Useful as the images were, however, no one could say with any certainty what kind of surface a lunar lander would encounter. The Moon's composition—which hinged in large part on where it came from— remained a mystery.

By the late 1960s, as the Apollo project neared its climax, scientists held a number of contending theories about the origin and composition of the Earth's only satellite. In one hypothesis, Earth and the Moon began 4.6 billion years ago as a single system, two lumps in the same primordial whirlpool of gas and dust. The second theory held that the Moon was created elsewhere in the Solar System, only to be captured later when it passed through Earth's gravitational field.

Neither of these explanations could account for a puzzling difference between the two bodies, however. From studying Earth's orbital dynamics, astronomers had deduced that the planet has a higher density than the Moon, a condition that is presumably the result of a heavy core of nickel and iron that formed when these elements sank inward to Earth's center; meanwhile, lighter minerals—made of elements such as silicon and oxygen—floated up toward the surface, forming the planet's outer layers, or mantle and crust. The Moon, in contrast, seemed to lack a heavy core and to be of more uniform density and composition throughout.

Using a special geologist's rake, Harrison Schmitt collects lunar pebbles at *Apollo 17*'s Taurus-Littrow landing site. Samples gathered during Apollo missions showed that the Moon contains minerals and elements that could support a lunar base.

A third pre-Apollo hypothesis attempted to reconcile these differences. Known as the fission theory, it suggested that Earth was created first. Then, some time after its core had formed, part of the lighter and still loosely compacted mantle material spun off and aggregated to form the Moon. However, the Earth would have to have been spinning unusually fast to throw off so much rock, and scientists were hard put to explain where the planet would have acquired this angular momentum.

As soon as Neil Armstrong and Buzz Aldrin crunched down off their ladder onto the Moon in front of a worldwide television audience in 1969, the process of resolving the issue began. Ultimately, twelve astronauts, including one geologist, walked on the Moon, sampling rocks and soil from six different regions near the Moon's equator and eventually returning about 840 pounds of Moon to the Earth.

Laboratory analysis of the moonrocks showed that some 40 percent of lunar soil is composed of oxygen bound up in mineral compounds. Silicon accounts for another 20 percent of the content of these compounds, and there are significant concentrations of iron, calcium, aluminum, titanium, and magnesium as well. The assays also revealed the presence of trace quantities of the lightest elements—hydrogen and helium—that had been deposited in the lunar soil by the solar wind, a powerful stream of charged particles flowing from the Sun.

Although the samples contained the same elements as Earth's mantle, albeit in different proportions, they were lacking in volatiles—water and gases that are easily driven off by heating. This supported another hypothesis for the satellite's origin. Proposed by planetary scientist William Hartmann, among others, the theory suggested that the fissioning of Earth and the Moon had been triggered by an enormous asteroid impact. If a massive body had crashed into the Earth just after the nickel-iron core formed, the resulting explosion would have blasted tons of light minerals, but no heavy ones, out of the mantle into space to become raw lunar material. The heat generated by the blast would also have cooked off the volatiles.

As promising as the impact theory is, it has yet to be proved, leaving the Moon's origins still cloaked in mystery. Other questions also remain. For example, since the Apollo landings were restricted to a belt near the equator, geologists learned nothing about the soil composition at the poles. Some scientists suspect that significant amounts of water ice might be frozen in the interiors of polar craters because the Moon's axis of rotation is nearly perpendicular to the Sun's rays: It thus has no seasons, and parts of the poles have never been exposed to sunlight. (One hypothesis is that this ice could have been formed from water vapor deposited by meteorites or comets that bombarded the Moon.)

Despite their limitations, the Apollo missions were an information bonanza. Discovering that the lunar surface contained useful metals as well as oxygen and hydrogen raised scientists' hopes for sustaining a human presence in space. Metals can be mined and refined into building materials; oxygen and hydrogen are major components of rocket fuel. They also combine to form water. Unless the poles turn out to be a hidden cache of water ice, that vital substance in its natural state may be entirely absent from the Moon. Extracting the ingredients for water from the lunar surface would be one way to provide water supplies to long-term human occupants. Armed with Project Apollo's legacy, space planners began to revive the old dream of building lunar colonies.

A MASS DRIVER

One visionary who had been thinking along those lines since the 1960s was Princeton physicist Gerard O'Neill, who began to study ways to use lunar products to build large structures in space. He also expanded upon Arthur Clarke's idea of an accelerator to launch objects from the lunar surface. Through the early 1970s, O'Neill pushed for the construction of ambitious orbiting space colonies, to be built primarily from materials mined from the Moon and shot into space by an accelerator, which O'Neill dubbed a "mass driver." The concept of the mass driver was simple. Lunar soil would be compacted into twenty-pound pellets and loaded into buckets. Each bucket would contain a set of coils made of a superconducting material that would create a powerful magnetic field; the coils would interact with a five-and-a-half-mile-long superconducting guideway beneath to give the bucket a friction-free ride *(pages 74-75)*.

A total of 300 buckets shooting one after another down the rail could launch a steady stream of materials spaceward at the rate of 50,000 tons of compressed moondust a month, or 600,000 tons a year. A self-propelled "catcher's mitt" stationed at a libration point—a stable gravitational point in the Earth-Moon system—would capture the pellets and funnel them into a storage bin. A space tug would periodically empty the bin and deliver its contents to a flying construction site, where the pellets would be smelted into their useful components: aluminum for the colony's superstructure, silicon for the glass panels, and oxygen for fuel and atmosphere.

Eventually, O'Neill wrote, space prospectors might acquire materials that are rare on the Moon or missing altogether, such as hydrogen, nitrogen, and carbon, by capturing asteroids and ferrying them to a habitat for mining. An asteroid would be moved by means of a solar-powered mass driver operating as a reaction engine. Attached to the asteroid by cables, the engine would eject pieces of it at high speeds, with the recoil pushing the combined mass of the asteroid and engine through space.

DRUMMING UP SUPPORT
Beginning in 1974, O'Neill organized Princeton conferences on space colonization and resource use that attracted wide media attention and helped persuade a cautious NASA to fund several low-cost studies on electromagnetic accelerators. O'Neill went on to found the nonprofit Space Studies Institute (SSI) to coordinate research into space manufacturing and the construction of space habitats. Between 1977 and 1983, SSI designed and built two prototype mass drivers; the second was able to move material at higher speeds along a shorter track than its predecessor. The research by SSI proved the feasibility of using mass drivers and laid the foundation for placing a full-scale version on the Moon. NASA went on to consider an altered form of the lunar mining scenario that O'Neill had popularized. In NASA's version, the proposed lunar base would include a mine and plant for processing liquid oxygen only; the propellant might eventually be shot into space with mass drivers.

NASA also conducted research through the 1960s on nuclear-

CREATING A WORLD WITHIN A WORLD

Thirty miles north of Tucson, Arizona, on the edge of the Sonora Desert, eight scientific pioneers are establishing the groundwork for human settlements on the Moon and beyond. To explore how future generations might survive in planetary colonies, these researchers have volunteered to be sealed up for two years, beginning in 1990, inside a massive greenhouse-like structure known as Biosphere II. (The Earth itself is Biosphere I.)

The inhabitants of Biosphere II, a privately funded experiment, will maintain communication with the outside world but will otherwise be entirely self-sufficient. Colonists will grow their own food in elaborate greenhouses. Carbon dioxide exhaled by humans and animals will help to nourish a wealth of plant life, which will in turn produce oxygen. Chemical and bacterial processes will purify organic wastes for reuse.

As shown at right, the closed ecosystem of Biosphere II will contain seven different biomes, or ecological zones: farmland *(1)*, a human habitat *(2)*, rain forest *(3)*, desert *(4)*, savanna *(5)*, marshes *(6)*, and even a miniature ocean thirty-five feet deep *(7)*. The 3,800 species of plants, insects, birds, fish, and small mammals that will share the 2.5 acres inside this glass-enclosed world will be carefully chosen to preserve ecological stability.

Biosphere II is designed to last for 100 years; the first two-year habitation may be followed by many longer-term experiments as humanity prepares to move outward into space.

COPY

thermal rockets. The test vehicles, propelled by hydrogen heated in uranium-fueled nuclear reactors, were slower but far more efficient than chemically powered spacecraft. In NASA's view, they could be suitable primarily for shuttling crews and cargoes between the orbits of Earth and the Moon in support of a lunar base. The experience gained by the Apollo astronauts showed that a permanent base was possible. However, without the focus of a dramatic, near-term objective like Apollo's goal of landing an astronaut on the Moon—accomplished in just eight years following President John F. Kennedy's 1961 announcement of the project—the manned space program lost much of its momentum. Distracted by economic problems that plagued the country through most of the 1970s, Americans questioned the wisdom of committing the huge amounts of money needed for such a venture. The government scrubbed the last three Apollo missions of the 1970s, and canceled the development of the nuclear rocket and other technologies required for establishing and maintaining a long-term human presence in space. In the end, NASA limited planning to activities that could be done in low Earth orbit with the upcoming space shuttle.

AN AGENDA FOR SPACE

Frustrated by the slowdown, many scientists and engineers both in and out of NASA began to press Ronald Reagan's administration and the Congress in the 1980s for a commitment to a more ambitious long-range space program. Scientists at NASA's Johnson Space Center in Houston began organizing workshops and conferences looking at specific long-term space projects. In 1985, Reagan appointed a National Commission on Space (NCOS) to recommend goals for the American space effort over the next half-century. O'Neill was one of the fifteen members of the commission, whose final report—ringingly titled "Pioneering the Space Frontier"—reflected both in spirit and in detail much of his once-unconventional thinking. The report called for an international space effort that would broadly advance science and industry, with a strong emphasis on human exploration of the Solar System and eventual settlement. It even included industrial spaceports at libration points (opposite).

America's movement into space would be achieved by "bootstrapping," said the authors—that is, by advancing in incremental steps, each building on the one before. This progress could start with establishment of the space station Freedom in the 1990s. The commission estimated the annual cost of a sustained, fifty-year commitment, including a permanent Mars base, at less than one-half of the amount spent on space activities during the apex of the Apollo program.

But unforeseen events overtook the commission's report. By the time it came out, in the spring of 1986, its impact had been blunted by the tragedy of the shuttle *Challenger* explosion. Spurred by the catastrophe, NASA appointed an internal task group, headed by astronaut Sally Ride, to examine the space agency's long-term direction and to chart possible strategies by

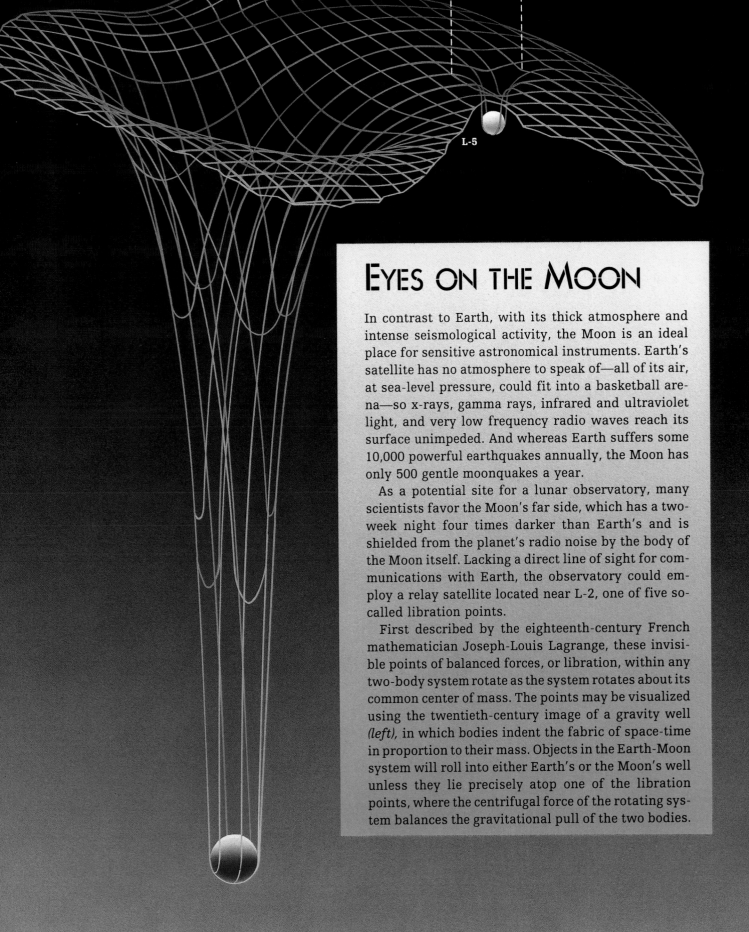

L-5

EYES ON THE MOON

In contrast to Earth, with its thick atmosphere and intense seismological activity, the Moon is an ideal place for sensitive astronomical instruments. Earth's satellite has no atmosphere to speak of—all of its air, at sea-level pressure, could fit into a basketball arena—so x-rays, gamma rays, infrared and ultraviolet light, and very low frequency radio waves reach its surface unimpeded. And whereas Earth suffers some 10,000 powerful earthquakes annually, the Moon has only 500 gentle moonquakes a year.

As a potential site for a lunar observatory, many scientists favor the Moon's far side, which has a two-week night four times darker than Earth's and is shielded from the planet's radio noise by the body of the Moon itself. Lacking a direct line of sight for communications with Earth, the observatory could employ a relay satellite located near L-2, one of five so-called libration points.

First described by the eighteenth-century French mathematician Joseph-Louis Lagrange, these invisible points of balanced forces, or libration, within any two-body system rotate as the system rotates about its common center of mass. The points may be visualized using the twentieth-century image of a gravity well *(left)*, in which bodies indent the fabric of space-time in proportion to their mass. Objects in the Earth-Moon system will roll into either Earth's or the Moon's well unless they lie precisely atop one of the libration points, where the centrifugal force of the rotating system balances the gravitational pull of the two bodies.

Bouncing a Message to Earth

Astronomers mulling over sites for a lunar observatory have considered a number of possibilities. One is on the far side near the limb, the line of demarcation between the Moon's far and near hemispheres. An observatory in this region could combine the advantages of far-side viewing with direct communications to Earth, since cables or relays could link the observatory with a base on the near side. Another site, in a permanently shadowed crater at one of the lunar poles, would be ideal for infrared systems and other instruments that need constant cooling.

The likeliest choice, however, is a site on the far side, near the equator. Shielded by some 2,000 miles of the Moon's bulk, an equatorial observatory would have a sweeping, full-sky view of the universe over the course of the Moon's cycle. And in an emergency, a transfer vehicle leaving from that latitude could easily intersect the orbital plane of a space station or other craft circling the Moon.

Far-side communications between the Moon and Earth could be handled by a relay satellite in a so-called halo orbit around the libration point L-2 *(below)*. Such a satellite, which appears to circle behind the Moon as the Moon orbits the Earth, actually flies in formation with L-2 around the entire Earth-Moon system, as illustrated at right.

Signals relayed off a communications satellite placed in a halo orbit some 40,000 miles behind the Moon around L-2 have a direct shot at Earth throughout the Moon's orbital cycle. The satellite can thus transmit voices and computer data, including digitized images of the universe from an equatorial observatory located on the Moon's far side.

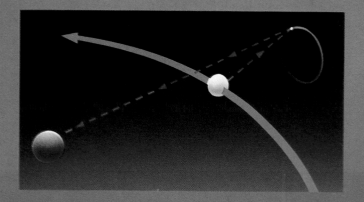

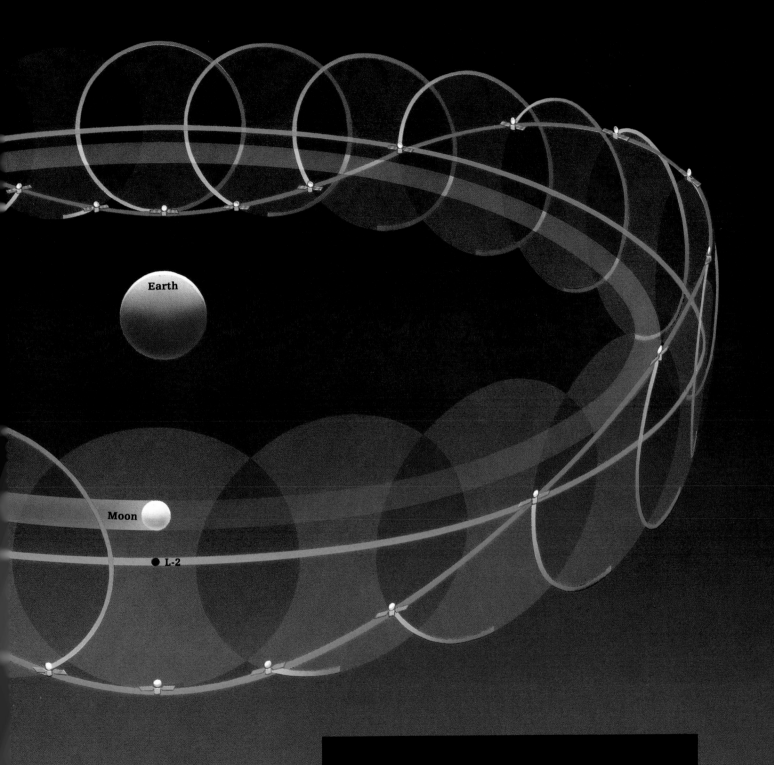

Because the Earth-Moon system's combined mass is unevenly distributed, the relay satellite's orbit *(pink)* as it tags along with L-2 *(black dot)* does not follow the same flat plane as the Moon's *(blue)*. Instead, it loops above and below the plane, making two apparent revolutions around L-2, and the Moon in front of it, for every orbit it makes around the whole system.

RAISING THE CURTAIN ON THE UNIVERSE

Located twenty-one degrees south of the lunar equator and ringed by mountains rising about two and a half miles above its flat floor, the 120-mile-wide Tsiolkovsky Crater *(right)* could hold a variety of instruments whose unique location would let scientists study more of the cosmos than ever. For instance, hundreds of radio antennas spread out in a vast array could allow them to detect for the first time the very low frequency radio waves blocked by Earth's ionosphere. A Moon-Earth radio interferometer could electronically connect lunar radio telescopes with those on Earth or in Earth orbit, yielding the resolution of a telescope with a diameter of 240,000 miles.

Liberated from structural constraints in the Moon's one-sixth gravity, engineers could build telescopes larger than any on Earth, with corresponding increases in resolution. Giant radio dishes might crouch like spiders over lunar craters, poised to gather the faint radio whispers that could signal the presence of extraterrestrial intelligence. X-ray and gamma ray detectors with collection areas at least two orders of magnitude greater than those of observatories planned for Earth orbit could pinpoint sources of radiation with unparalleled accuracy. In the optical range, a lunar array would provide resolution 100,000 times better than that of Earth-based telescopes and 10,000 times better than that of the Hubble Space Telescope. Such extraordinary instruments might capture the first images of other planetary systems— or the seething energy from black holes.

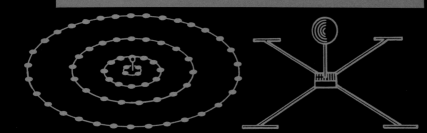

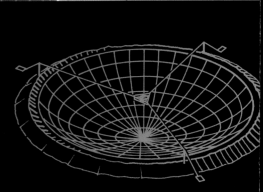

Among the many instruments planned for a lunar observatory are 200 to 300 three-foot dipole radio antennas dispersed over a circle about twelve miles wide. Clustered into groups of four, the antennas will be connected to a small box of electronics that transmits the data to a central correlator. This Very Low Frequency Array would detect radiation—such as low-energy particles from pulsars and active galaxies—that cannot penetrate Earth's electrically charged ionosphere, offering clues to the nature of extragalactic radio sources.

Searching for emissions from stars within eighty light-years of the Sun, 5,000-foot-wide radio dish antennas situated in lunar craters will canvass the Milky Way for signs of extraterrestrial intelligence and map the gas in it and other galaxies.

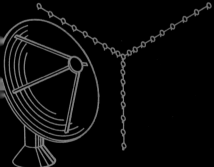

The Moon-Earth radio interferometer's 65-foot-wide dishes will form a Y with twelve-mile arms to simulate the aperture of one huge dish electronically linked to Earth. By examining distant galactic clusters, scientists hope to determine the universe's expansion rate.

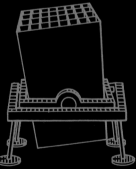

Collection areas thirty-five feet wide could enable lunar-based detectors for high-energy x-rays and gamma rays to provide insights into the formation of galactic clusters and the nature of emissions from gaseous accretion disks around black holes.

LOUISA, the Lunar Optical-Ultraviolet-Infrared Synthesis Array, will consist of forty-two domed telescopes arranged in two circles. The multipurpose array may be able to find dim planets around other stars or peer into the cores of elliptical galaxies.

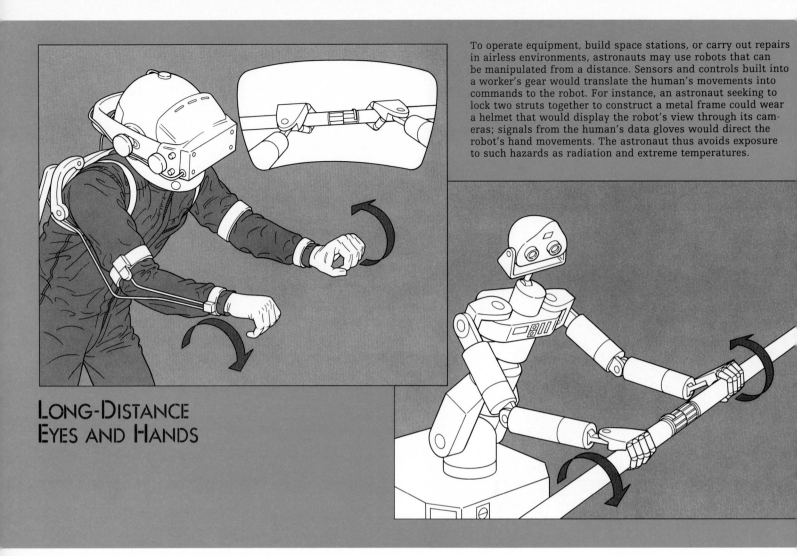

To operate equipment, build space stations, or carry out repairs in airless environments, astronauts may use robots that can be manipulated from a distance. Sensors and controls built into a worker's gear would translate the human's movements into commands to the robot. For instance, an astronaut seeking to lock two struts together to construct a metal frame could wear a helmet that would display the robot's view through its cameras; signals from the human's data gloves would direct the robot's hand movements. The astronaut thus avoids exposure to such hazards as radiation and extreme temperatures.

LONG-DISTANCE EYES AND HANDS

which the United States could regain its once-undisputed leadership in space.

The Ride report, which was published in August of 1987, laid out four potential initiatives for NASA. One would be a scientific study of Earth from the vantage point of space, focusing on such global problems as declining resources and atmospheric warming. Another scenario was exploration of the Solar System using unpiloted probes. A third option would lead to a permanent lunar base; a fourth provided for missions to Mars, ending in eventual human habitation.

While acknowledging the need to set priorities, the panelists on the Ride task force emphasized that it would be a mistake to select any one of these initiatives to the exclusion of the others. Studying Earth from space and sending probes to distant parts of the Solar System were equally valuable undertakings, they declared. However, they were skeptical of a one-shot mission to Mars, which could turn out to be a costly dead end rather than a solid basis for further exploration. A better idea, the panel concluded, would be to take an evolutionary approach, including the establishment of a permanent base on the Moon, which would represent a sustained commitment toward the goal of living and working in space. Even as it picked up where the Apollo project left off, a lunar station would be a logical stepping-

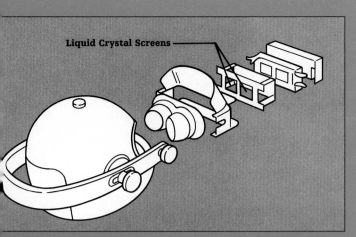

Liquid Crystal Screens

stronauts wishing to see through the camera eyes of a robotic oxy can don a special helmet containing two television screens that mbine for a three-dimensional image.

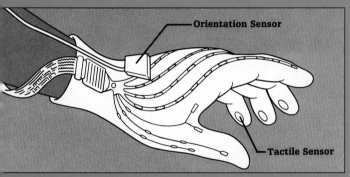

Orientation Sensor

Tactile Sensor

nsors on the palm, fingertips, and back of a so-called data glove ansmit to a robot the movement, gripping force, and position in ace of the human user's hand.

stone for sending astronauts to Mars and beyond.

Not everyone agreed with this conclusion. Various "Martians," as they were sometimes called, soon weighed in against the "Moon firsters" with arguments for a dash to the Red Planet. Some people within the NASA fold, including former Apollo astronaut Michael Collins, believed that only a national, Apollo-like venture of landing a human on Mars would get people excited about the space program again. Cornell astronomer and science-TV host Carl Sagan urged an all-out Mars effort and viewed a lunar base not as "a detour on the road to Mars, but a trap" that would use up limited money and delay a human Mars mission indefinitely. The Planetary Society, a national group of space enthusiasts of which Sagan was president, circulated a "Mars Declaration" urging humanity onward to Mars; Nobel laureate chemist Linus Pauling and former president Jimmy Carter, among others, signed it.

In June of 1987, NASA established the Office of Exploration (OEXP) under engineer and Apollo-era manager John Aaron to focus planning for manned voyages beyond Earth. The cause of space exploration gained administration support when on July 20, 1989, the twentieth anniversary of the first moonwalk, President George Bush announced his backing for both a Moon base and a Mars expedition. Although Bush did not specify deadlines or say how either program would be financed, his announcement gave heart to proponents of exploration. If the United States, possibly aided by international partners, was to make a firm commitment to it, a lunar base could be in place as early as 2001.

In the late 1990s, NASA should complete construction of the space station Freedom, while collecting information from a series of robotic lunar probes. Equipped with remote sensors to determine the composition of the lunar soil, detect the Moon's magnetic fields, and collect images of the Moon in different parts of the electromagnetic spectrum, a satellite known as the *Lunar Observer* will circle the Moon from pole to pole, mapping its geochemical features in exhaustive detail. In the *Observer*'s wake, robot prospectors may land on the Moon to assay sites for the creation of a lunar mining base—an enterprise that NASA's experts have imagined in detail. . . .

HOME ON THE MOON

. . . Soon after the turn of the century: NASA has modified Freedom to accommodate the construction facilities needed for servicing Moon-bound ships. The expanded station has a new module to house the full-time crew, along with a bay for maintaining the lunar spacecraft. Heavy-lift rockets have ferried the components and fuel from Earth to orbit, and workers floating in bulky spacesuits in a zero-gravity environment or guiding tele-robotic arms mate a transfer vehicle and cargo vehicle and fuel them. The lunar transfer vehicle, powered by conventional chemical propellant, is to carry crews between low Earth orbit and lunar orbit. The other, an unpiloted cargo ship, will carry within it a three-in-one vehicle to shuttle crew, cargo, and propellant from lunar orbit to the Moon's surface and back again. The cargo ship will rely on a small nuclear reactor to gene-rate electricity to accelerate ionized gas, providing thrust. Although this engine will yield less thrust than conventional rockets, it will save on precious rocket fuel.

Within weeks, the lunar spacecraft are assembled and ready to go. After the three-day trip to lunar orbit, the crew transfers to the cargo ship, where they enter the three-in-one lander and ferry cargo to the lunar surface, touching down in the Mare Tranquillitatis, near the site of the first Apollo landing some thirty years ago. There, the weary travelers face the unpleas-ant task of unpacking.

The first few stays are frustratingly short ones. Forced to depend on the limited oxygen supply brought from Earth, the four-person construction crews can stay only eight days—little time for the lunar golf popularized by Alan Shepard back in 1971 or for roaming the rugged mountains.

The habitat module and a first small oxygen mine and processing plant, which will produce 150 tons of liquid oxygen for propellant, are ready in the first year. The living structure, made of a high-strength, multi-ply fabric supported by an inner framework of light, strong graphite, is partially buried in the lunar surface; in its early phase of construction it resembles a giant beachball. The dome is about fifty feet in diameter and is subdivided into four floors, where the crew members live and work in a shirt-sleeve environment. On top of the building, astronaut workers and robotic helpers pile a nine-foot-thick layer of bags filled with lunar soil to protect the crew from radi-ation and micrometeoroids. The covering also insulates the habitat against the extremes of heat and cold.

From first landing to development of the base as a fully operational facil-ity takes about two years. Most of the heavy construction at the site— leveling, grading, and excavating—is done by a versatile machine called the prime mover, which, depending on its attachments, can function as a bulldozer, a front-end loader, or a backhoe. Like many of the other devices at the lunar outpost, the prime mover is remote-controlled and can be oper-ated either from a station in the living quarters or from Earth. Stereo cameras and other on-board sensors give the operator the hands-on feel of riding the machine *(pages 62-63)*.

As the cargo ship continues its runs, bringing more equipment and supplies, the base gradually takes shape. Two power stations—one nuclear, the other

One NASA design for a Moon base calls for an inflatable sphere, fifty-two feet wide and made from a tough cloth such as the kind used in bulletproof vests, partially buried in the lunar surface. Soil heaped around the perimeter of the exposed dome would protect its inhabitants from lethal radiation. Powered by nuclear and solar plants *(background)*, the habitat module could accommodate up to a dozen crew members.

solar—are placed in operation. The nuclear station produces continuous power, with the solar panels providing additional energy during the two-week-long lunar day.

LUNAR SCIENCE

Once the basic necessities of lunar life are provided for, the visiting astronauts increasingly turn their attention to scientific inquiry. One of the most important things they study is themselves. They run each other through a battery of tests on treadmills and stationary bicycles. They jump, sleep, and give blood and other bodily fluids to find out how the body works in the light lunar gravity and harsh environment.

In some ways, just being there is the biggest experiment of all. What happens to a handful of people in close quarters for a year on the Moon? The answers will serve NASA in planning the far more psychologically arduous journey to Mars and back, a voyage that could take nearly three years round-trip, including a year's stay on the Red Planet. NASA scientists have been studying the psychological effects of isolation and confinement since the 1960s. Some of their data has come from space stations and from crews at remote research bases in Antarctica, where people

A lunar tug ferries a multiton load from the Moon into space in this artist's rendition of a NASA proposal for a so-called transfer vehicle. Fueled in part by liquid oxygen extracted from moonrocks, such craft would transport cargo and passengers between lunar orbit and ships or space stations in the Earth-Moon system.

have spent winters since World War II. During the typical stay in Antarctica, which lasts between seven and nine months, even the most well-adjusted people suffer from loss of sleep, lack of energy and concentration, anxiety, and social withdrawal. Antarctic veterans joke about maladies like "big eye," characterized by circles under the eyes from lack of sleep, and "long eye," a vacant stare. On the Moon or halfway to Mars, such fatigue and stress could have serious consequences.

By the 1990s, the Soviets had kept cosmonauts in orbit for more than a year at a stretch, thus acquiring the most experience dealing with the psychological effects of space. Not surprisingly, they found that little things—allowing the astronauts time off, regular radio conversations with their families, and freedom to decorate their personal living space—can make a profound difference in morale.

Not all the work for the lunar pioneers is so introspective. They carry on the geophysical investigations begun by the Apollo astronauts—taking seismic data and extracting deep rock samples by drilling two-thirds of a mile into the lunar crust. Once in a while they send a remote-controlled rover out to cruise the lunar landscape, set up experiments, and retrieve samples from more than 100 miles away.

While work proceeds at Tranquillity base, another scientific mission unfolds several hundred miles away. There, on the side of the Moon permanently turned away from the Earth, astronauts construct an astronomical observatory. A pair of cargo missions deliver an array of instruments that includes x-ray, infrared, visible-light, and radio telescopes, and visiting four-person crews deploy them. Thereafter, the observatory is visited every three years or so by astronauts from Earth orbit to maintain the instruments and make repairs. The instruments themselves are operated remotely by radio or laser from Earth, and data is transmitted back the same way.

The Moon offers several advantages over Earth for observing the cosmos. Most important, it has no blocking atmosphere and, on the far side, no radio interference. In this favorable environment, the various instruments—electronically connected to some on Earth—are capable of many times the resolving power of any terrestrial counterpart and can probe the distant reaches of the universe *(page 57)*.

LUNAR INDUSTRY

Back on the near side, lunar astronauts proceed to develop the Moon's potential as a technological test bed and as part of the infrastructure for deeper space travel. The astronauts will eventually guide machines that extract and refine useful materials—such as hydrogen, helium-3, aluminum, and titanium—from the lunar soil. But the first product of the mining base is the essential element oxygen.

The oxygen-processing plant borrows from the technology of terrestrial open-pit mining and is almost completely automated, requiring only one astronaut to supervise its operation. A robot-controlled front-end loader scoops up fragmented lunar rock and drops it into a screening device called a scalper. First the chunks of rock pass through a grate to separate out the large pieces. Then the smaller rocks travel through crushers and are ground into a coarse-grained feedstock.

A superconducting electromagnet pulls grains of ilmenite, a magnetic oxide, from the feedstock, and a reactor heats them to 1,800 degrees Fahrenheit in the presence of hydrogen. At that temperature, oxygen atoms can be stolen away from the oxide by hydrogen atoms, forming water. Electrodes break down the water molecules into H_2 and O_2. The oxygen is liquefied and the hydrogen recycled to be used again in the extraction process.

The oxygen from the pilot plant goes into propellant and gases for breathing. Once engineers have thoroughly tested the techniques for processing lunar soil, the plant will be expanded, and most of the oxygen produced will be liquefied and stored for use as rocket propellant. When the base is geared up to extract lunar hydrogen, propellants from the Moon could be used for Mars missions. Lunar freighters would carry tanks of oxygen and hydrogen up to a cargo ship waiting in lunar orbit, and the ship would ferry the oxygen back to the space station Freedom, where the Mars rocket would be under construction.

Once a base exists on Mars and interplanetary trips become frequent, the demand for rocket fuel will necessitate finding a more efficient way of getting the liquid fuel into space. At that time, lunar residents may expand their mining operations and build a mass driver to catapult the liquid off the lunar surface and into orbit. In effect, the astronauts and their robots will be running a long-distance pumping station. . . .

LUNAR LIVING

. . . All of this is speculation, but it is based on hardheaded studies by some 200 physicists, astronomers, and engineers working under the auspices of NASA's Office of Exploration—and that research, in turn, draws on decades of prior work conducted by scientists during the ambitious days of the Apollo era. Beyond NASA's planning and detailed guesswork, the picture becomes blurred. Perhaps, as the Moon slowly becomes domesticated, earthbound residents will come to regard traveling there as routine. Returning astronauts will no longer automatically be invited to visit heads of state; international crews will become more diverse and include journalists and artists. Whether the Moon will be colonized in the conventional sense of supporting large numbers of people living, working, growing their own food, and raising families remains uncertain. But the pioneers who take their turns of duty there, exploring the dun-colored ridges and ravines, supervising the silent grind of the mining machines, turning their eyes toward the unwavering crowd of stars, will know they are vital participants in a grand odyssey.

THE MOON'S RESOURCES

The establishment of a permanent human presence in space will depend on efficient and reliable supply lines. However, the cost of launching a continual stream of carriers from Earth would quickly become prohibitive. To break free of the planet's gravitational clutch and go to the Moon or beyond, vehicles must accelerate to an escape velocity of about 25,000 miles per hour, a speed that can require the expenditure of hundreds of thousands of gallons of rocket propellants such as liquid oxygen and liquid hydrogen. The heavier the payload, the greater the amount of fuel needed simply to launch a vehicle into Earth orbit as well as to fight the drag created by Earth's thick atmosphere. For interplanetary missions to be more practical, then, space engineers must reduce reliance on the home planet—by finding off-Earth sources of building materials, for example, and of fuel. Even if food and goods continued to be shipped into space from Earth, eliminating the need to boost fuel for outbound vehicles off Earth's surface would significantly cut the cost of supplying distant posts.

As illustrated on the following pages, a mining facility on the Moon would be one step in loosening ties to Earth. The lunar surface is a potentially rich source of metals and minerals for building materials. It is also rich in oxygen, which could be processed into liquid propellant and oxygen for life support. The map above shows some of the sites under consideration for an outpost with lunar oxygen mining; all are near the equator for easy rendezvousing with orbiting transfer vehicles or space stations. Because the Moon has virtually no atmosphere and only one-sixth Earth's gravity, the amount of fuel required to escape the lunar surface is a fraction of that needed on Earth.

A LUNAR MINING OPERATION

Bound up with iron, calcium, aluminum, and other elements in moonrock mineral compounds, oxygen accounts for about 40 percent by weight of the material making up the Moon's soil. Extracting it from its mineral host requires breaking its chemical bonds with other elements. In some common lunar minerals, oxygen content is high, but these bonds are very strong, making extraction difficult and expensive. By contrast, certain oxygen bonds of an iron-titanium-oxygen mineral called ilmenite are more easily broken. In addition, the mineral itself can be found in relative abundance on the lunar surface. On Earth, ilmenite is a source of the titanium used for pigment in paint and other surface coatings; on the Moon, it constitutes as much as 25 percent by volume of basaltic rock in the sealike lunar maria, areas where meteorite impacts dug great craters and blasted up ilmenite-rich basalt.

Oxygen mining is likely to begin in these cratered regions, where the high ilmenite concentrations may offset the labor involved in the intensive grinding and crushing and other steps necessary to prepare the basalt for oxygen extraction. Once it is operating at full capacity, a lunar oxygen facility like the one pictured here could mine about ten acres a year. Telerobotic excavators and haulers—directed remotely by human operators—could process perhaps 185,000 tons of lunar basaltic rock annually, separating some 1,000 tons of liquid oxygen.

From the mining site (1), telerobotic haulers carry basaltic rock particles no larger than ten inches in diameter to the processing plant (2), where the rock is put through three stages of crushing and grinding. Reduced to a fine powder, the basalt goes into magnetic separators (3 and inset, far right), which winnow out grains of iron-bearing ilmenite in preparation for the oxygen extraction reactor (4 and pages 72-73). After being refrigerated, liquid oxygen is stored (5). Waste material is then hauled back to a dumpsite (6).

Electric coils in the cylindrical walls of a magnetic separator generate an electromagnetic field that draws ilmenite particles away from other material in the pulverized rock. Nonmagnetic rock particles fall straight through the cylinder into a waste collection pipe. Ilmenite, because of its iron content, is attracted toward the magnetized walls and falls outside the waste pipe.

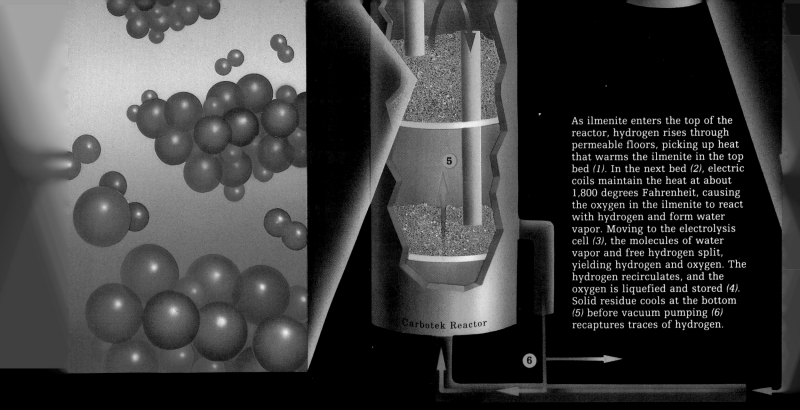

As ilmenite enters the top of the reactor, hydrogen rises through permeable floors, picking up heat that warms the ilmenite in the top bed *(1)*. In the next bed *(2)*, electric coils maintain the heat at about 1,800 degrees Fahrenheit, causing the oxygen in the ilmenite to react with hydrogen and form water vapor. Moving to the electrolysis cell *(3)*, the molecules of water vapor and free hydrogen split, yielding hydrogen and oxygen. The hydrogen recirculates, and the oxygen is liquefied and stored *(4)*. Solid residue cools at the bottom *(5)* before vacuum pumping *(6)* recaptures traces of hydrogen.

Carbotek Reactor

SNATCHING OXYGEN FROM ILMENITE

One of the many methods proposed for extracting oxygen from moonrock is a procedure—based on a process developed by the Texas engineering firm Carbotek Incorporated—that involves using hydrogen to react with ilmenite powder. (Initially, liquid hydrogen would be shipped up from Earth; in the longer term, hydrogen could also be recovered from lunar soil for use in the reactor and for fuel.) During the process,

virtually all of the hydrogen would be recycled in a triple-decker reactor like the schematic one above.

Because the bonds between the oxygen and iron atoms in ilmenite are relatively weak, hydrogen molecules are able to make new bonds with oxygen to form water vapor. Water molecules are like little bipolar magnets, with a slightly negatively charged oxygen atom at one end and positively charged hydrogen atoms at the other. When placed in an electrically conducting medium such as zirconia and electrified with a current, the water molecules break up. The now-separated oxygen is cooled to about −150 degrees Fahrenheit and stored under pressure in liquid form,

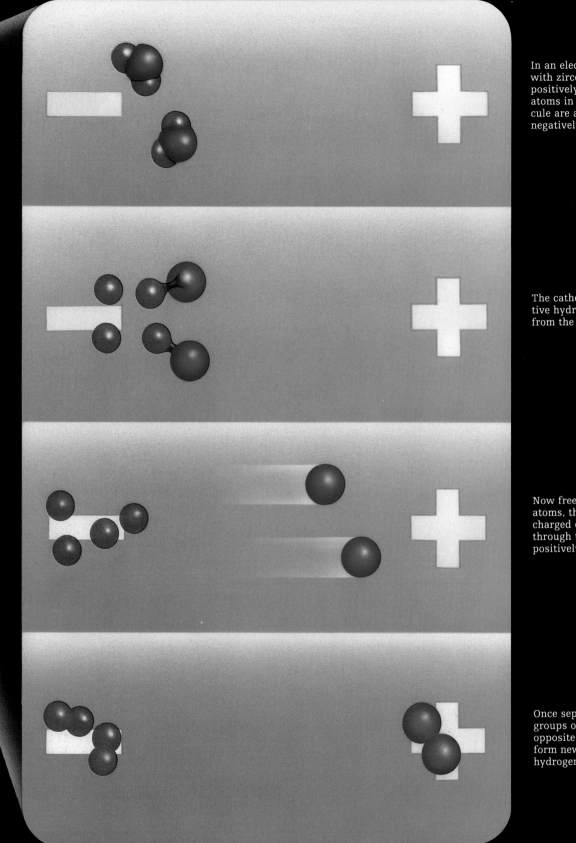

In an electrolysis cell filled with zirconia, the two positively charged hydrogen atoms in each water molecule are attracted to the negatively charged cathode.

The cathode pulls the positive hydrogen atoms away from the water molecule.

Now free of the hydrogen atoms, the negatively charged oxygen atoms flow through the zirconia to the positively charged anode.

Once separated, the two groups of atoms combine at opposite poles of the cell to form new molecules of hydrogen (H_2) and oxygen (O_2)

The magnetic fields generated by the payload bucket coils *(red)* and the stationary series of drive coils *(blue)* share the same orientation of their north and south poles. When the electric current in the drive coils is switched on and off in a carefully timed sequence, attraction between the north pole of the bucket coils and the south pole of the drive coils pulls the bucket forward, accelerating it to escape velocity.

N — S   N — S

SLINGSHOT

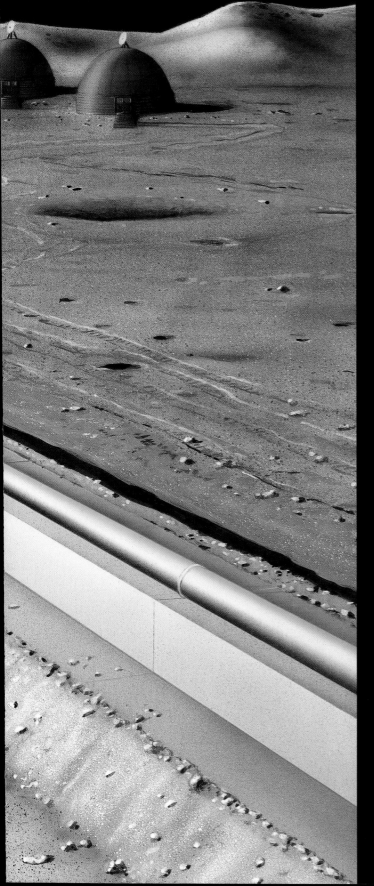

Space engineers have envisioned a number of methods for transporting liquid oxygen from a lunar mining facility to orbiting fueling stations. One scenario calls for a spacefaring propellant tanker that would make periodic landings at the processing plant and ferry fifty-ton payloads of liquid oxygen into lunar orbit, there to be transferred to vehicles outbound from Earth. Such a tanker would employ a conventional rocket engine and burn a propellant mixture of liquid hydrogen and oxygen. In the weak lunar gravity, the tanker would require scarcely one-twentieth the energy needed to launch from Earth.

An even more energy-efficient possibility, illustrated at left, is an electromagnetic mass driver, a device that makes use of the force of superconducting magnetic attraction to propel an object down a 500-foot tube. Liquid oxygen could be placed in bucket-shaped containers, each weighing perhaps 150 pounds loaded and equipped with external electric coils; a current flowing through the coils would generate a magnetic field. A mass driver could then accelerate a series of these buckets to the necessary speed, lofting them into space while burning a minimum of fuel.

Riding on eddies of magnetic force, buckets with liquid oxygen accelerate through a long tunnel formed by a series of hoop-shaped drive coils *(inset)* in this rendition of a mass driver at a lunar mining facility. When they reach escape velocity—almost 5,400 miles per hour, or 7,875 feet per second—the buckets shoot out into space. According to some scenarios, the buckets might be, in effect, little robots, programmed to deliver themselves to a station in orbit around the Moon or Earth where spacecraft bound for other destinations could refuel.

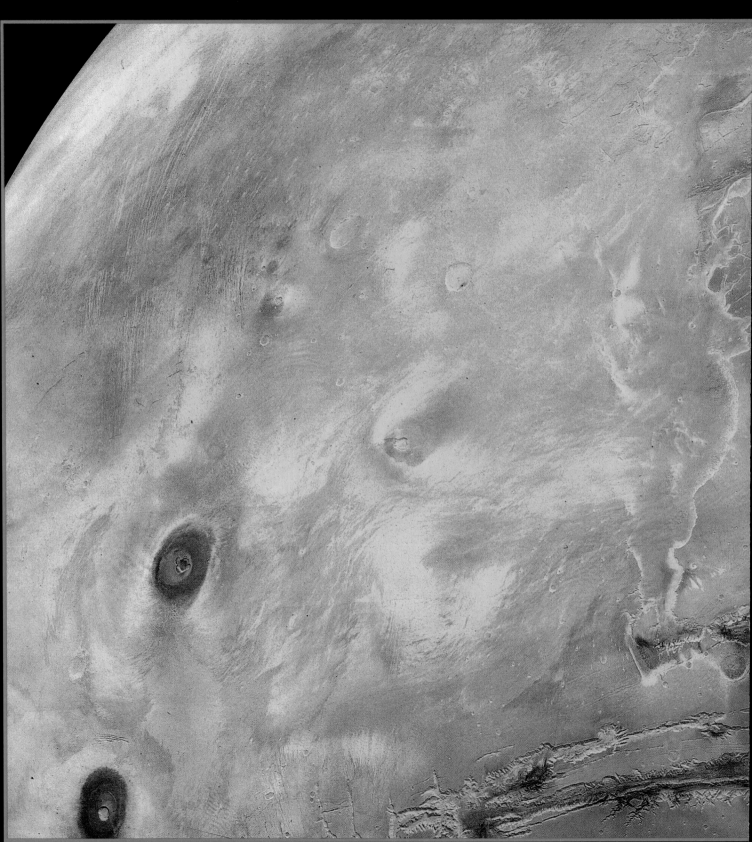

land near Valles Marineris, the two-mile-deep canyon that runs for 2,800 miles along the planet' equator. The layered walls of its tributary valleys will give geolo- gists clues to Mars's past.

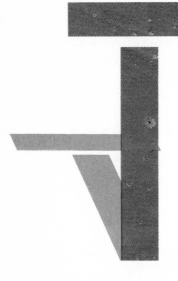

wo geologists work intently in a small laboratory. One of them glances up from her instrument display to check the clock. Judging that her experiment can manage itself for a few minutes, she heads toward an open doorway. Once an air lock that led outside, it is now an entrance to the transparent dome of a greenhouse. Her husband, a botanist farmer, awaits her there for a late-afternoon stroll among neat rows of wheat and beans and past a teeming fishpond.

With sunset only minutes away, the geologist gazes intently across the ocher-colored sand dunes that have become an all-too-familiar landscape. Her four-year stay at the Mare Boreum base near Mars's north pole has not been easy, nor has it been without hazard. That golden sunlight, for example, belies the temperature outside the double-walled protective enclosure—sixty-five degrees below zero Fahrenheit and plunging. Wisps of haze overhead give no clue that the atmospheric pressure is a scant seven millibars—which on Earth corresponds to an altitude of more than 100,000 feet—or that this rarefied mix of gases consists almost entirely of carbon dioxide. A human being stepping outside without the protection of a spacesuit would expire within seconds.

Still, Mars has its attractive features. For example, the planet's day, called a sol, is only 37.4 minutes longer than Earth's, so the daily routine of sleep and work seems perfectly normal. Another advantage is the weak gravity, only three-eighths that of Earth, which makes light work of heavy labor. Then, there is the predictably sunny sky, which never bodes the slightest chance of precipitation.

The geologist catches sight of a bright pinpoint of light emerging from the quickly deepening dusk. Earth's blue-tinged beacon is unmistakable. In a few days, a space ferry will carry one-half of the base's twenty-four inhabitants, including the geologist and her husband, to their home planet. Though the ferry is fitted with powerful rockets to speed the journey across 200 million miles of space, the trip will take four months, a long time to pioneers looking forward to reunions with family and friends. . . .

AN OLD URGE
. . . Any colonization of Mars by venturesome earthlings is unlikely to occur before the end of the twenty-first century, yet the roots of the idea can be traced back to the very early years of the twentieth century, when Konstantin

Tsiolkovsky painstakingly worked out the details of a multistage "rocket train" that could escape the gravitational bounds of Earth. Reflecting in 1911 on his earlier breakthroughs, he wrote, "Mankind will not forever remain on Earth, but in the pursuit of light and space will first timidly emerge from the bounds of the atmosphere, and then advance until he has conquered the whole of circumsolar space."

In 1920, the Russian rocket experimenter Friedrich Tsander, a contemporary of Robert Goddard in the United States and Hermann Oberth in Germany, met Vladimir Ilyich Lenin after delivering a lecture on spaceflight in Moscow. During his encounter with the revolutionary leader, Tsander mentioned his research into the means by which human beings might rocket to Mars—including suitable apparel for the journey, a nutritious diet, and ways for spacefarers to withstand the rigors of acceleration. "Lenin shook my hand strongly," noted Tsander in his diary, "wished me success in my work, and promised support."

None was forthcoming. In the Soviet Union, as in most other nations, governments had more pressing concerns than converting science fiction to fact. Thoughts of Mars and the nascent field of rocketry were left to amateurs, who often formed organizations such as the Group for the Study of Reactive Motion, founded by Tsander and several other like-minded Muscovites in October 1931. Among the members of this association of enthusiasts—whose meetings were punctuated with shouts of "Onward to Mars!"—was Sergei Korolyov. Two and a half decades later, Korolyov gained renown in the Soviet Union as the engineering mastermind who spearheaded his country's spectacular entry into the space age with the orbiting of *Sputnik 1* on October 4, 1957.

PROGRESS

Lofting a satellite into Earth orbit was a small but encouraging step toward the kind of Mars mission that Germany's Wernher von Braun had outlined five years earlier. Drawing on his wartime experience with the V-2 missile—as well as his later work in the United States along similar lines—von Braun produced the first comprehensive scenario for reaching Mars. In 1952, the magazine *Weltraumfahrt,* German for "space travel," published a special issue devoted to *Das Marsprojekt,* the rocket engineer's plan for the exploration of Mars by "a flotilla of ten space vessels manned by not less than 70 men." The massive spacecraft he envisioned would be assembled in orbit from components ferried up from Earth. They would be fueled by hydrazine and nitric acid, well-known propellants in the rocket trade. He eschewed the use of more powerful mixtures like liquid hydrogen and oxygen, believing that their low temperatures made them too difficult to handle and store, and he argued against untried means of propulsion—such as nuclear power— as too costly to develop.

Von Braun's assault on the fourth planet would rely on energy-efficient interplanetary trajectories that had been discovered in 1925 by Walter

One NASA proposal for Mars exploration at the turn of the century calls for a sophisticated multistage probe to go into Mars orbit and deploy a lander to the surface. Equipped with a robotic arm, the lander would scoop up samples of soil and rock. Then, using its bottom portion as a launch pad, it would lift off for a rendezvous with the return stage of the orbiter—one of two craft shown passing overhead in this artist's rendition of the mission. The samples would be transferred to the return vehicle, which would take them back to Earth for analysis, leaving the carrier stage—which ferried the lander to Mars—to remain in orbit.

Hohmann, a German architect and engineer. Hohmann realized that the most fuel-efficient routes between planets would take advantage of the Sun's gravity to shape elliptical orbits much like the trajectories that asteroids and comets follow. Named for their inventor, these routes are known as Hohmann transfer orbits. If one end of a spacecraft's elliptical path just touches the orbit of Earth and the other end coincides with the orbit of the destination planet, the ship expends the least energy possible to get there *(page 89)*. To achieve this result, it is necessary only to time the launch so that the target planet and the spacecraft come together at the point where their orbits intersect.

One disadvantage of a Hohmann route to Mars is that it makes for a long journey. The average distance between Earth and Mars at opposition—the moment every 780 days when the two planets line up on the same side of the Sun and are therefore at their closest—is 48 million miles. However, because the orbit of Mars is much more elliptical than Earth's, the actual distance for a given opposition can vary by as much as 30 million miles. So-called perfect oppositions, when the orbital movements of the two planets shrink the distance between them to just 35 million miles, occur only once every 284 years, although a pair of good oppositions occur sometime near the month of August every 15 and 17 years.

The length of time it takes a spacecraft to travel a Hohmann trajectory to Mars will therefore vary by as much as six months, depending on exactly when it is launched. By von Braun's calculations, 270 days would pass between the moment that astronauts blasted out of Earth orbit and their arrival in orbit around Mars. Shorter, faster routes are possible, but they have the drawback of requiring more fuel—both to speed up at the outset and to slow down upon reaching the destination.

A LUNAR STEPPINGSTONE

Prospects for space travel were given a huge boost on May 25, 1961, when President Kennedy announced that the United States intended to send an expedition to the Moon within the decade. Von Braun and others in the fledgling National Aeronautics and Space Administration quickly became enthusiasts of Kennedy's Project Apollo; such a mission offered an opportunity to develop the hardware required for flights deeper into space, and Mars was the obvious next step.

Indeed, even as the Moon-bound Apollo vehicles were taking shape in the spring of 1962, hundreds of NASA engineers in adjacent buildings—as well as their counterparts at aerospace contractors far and wide—were spending millions of dollars planning future trips to Mars and speaking confidently of their schemes for other interplanetary missions. At one meeting in November 1964, for instance, engineers announced that they could land Americans on Mars just nine years after getting the go-ahead for such a mission. But their spacefaring dreams were simply not affordable. Instead, NASA committed its scarce funds to more modest research efforts, such as the Earth-orbiting space station called Skylab.

Despite the lack of official encouragement, Wernher von Braun was evolving a new, improved Mars Project. Considerably scaled back from his 1952 plan, it was still an ambitious proposal. Two spaceships would leave for Mars after being assembled in Earth orbit. The vessels, each with a crew of six, were to be powered by nuclear engines using hydrogen as a propellant—the very ideas that von Braun had rejected nearly two decades earlier. He could embrace hydrogen because techniques for handling and storing the gas in liquid form had improved in the interim. As for nuclear power, tests in the Nevada desert had already confirmed the practicality of such an engine.

During the 1950s and 1960s, Project Rover, a joint endeavor by the Atomic Energy Commission, the Air Force, and NASA to create a nuclear-powered rocket engine, had developed a number of variations on a common design theme: replacing the combustion chamber of a conventional chemical rocket with a nuclear reactor. The effort relied on fission, the power behind the first atomic bombs and all present-day nuclear power plants. In a fission reaction, the nuclei of atoms of heavy elements such as uranium are bombarded by neutrons and split, leaving behind highly radioactive lighter nuclei but also converting a small portion of their original mass into a vast amount of energy in the form of heat.

Hydrogen propellant, stored in liquid form aboard the spacecraft, would be heated to 4,000 degrees Kelvin by piping it through the core of the reac-

tor. It would streak from the spacecraft's tail at six miles per second—about twice as fast as the exhaust of the most powerful chemical engine possible—accelerating the vessel to nearly the same speed. Such an engine would require less fuel to reach Mars, reducing the cost of the mission, though not its duration.

As von Braun's two spaceships orbited the planet, half the crew members would descend for a firsthand look at the Martian surface. After eighty days on the ground, the astronauts would rejoin the mother ship and head for home by way of Venus, stopping there long enough to drop off instrumented capsules and probe the planet's cloud-shrouded surface from orbit with powerful radar systems. By the time they returned to Earth, the crew would have coursed through interplanetary space for some twenty-one months in all.

The scheme died aborning. Scarcely a month after von Braun presented his plan to the U.S. Senate, a panel called the Space Task Force, created by President Richard Nixon in early 1969 to recommend post-Apollo undertakings for NASA, delivered its report. Chaired by Vice President Spiro Agnew, the task force pitted eager proponents of space exploration like NASA administrator Thomas Paine against cost-focused officials like Robert Mayo of the Bureau of the Budget. In the end, while Agnew personally supported a vigorous exploration program, the report painted the idea of astronauts on Mars as an extremely ambitious option and buried it in rhetoric that spoke of "a balanced program," "flexibility in options," and of course, "severe budget constraints." Without explicitly directing its deferment, the Space Task Force effectively tabled human exploration of Mars until well into the twenty-first century—at the earliest.

THE VIEW FROM MARINER

The panel was much more enthusiastic about sending unpiloted probes to the inner planets of the Solar System. Some four years earlier, for example, the American spacecraft *Mariner 4* had flown within 8,000 miles of Mars, radioing back to Earth twenty-two television images, the clearest views yet of a world steeped in legend and speculation. The planet that *Mariner 4* revealed was a far cry from the romantic vision of a world with cities linked by networks of canals. Instead, Mars's desolate surface was pocked by hundreds of huge craters that attested to eons of battering by chunks of interplanetary debris. In many respects, the planet resembled Earth's moon. This

SIMULATED FLYOVERS

Pilots training for Mars landings will spend many hours on computer-simulated flights over the planet's surface, watching volcanic cones and vast canyons flash by on video screens as they descend. The realism those training flights will have is foreshadowed by the image at near left, one frame from a five-minute film called "Mars, the Movie," generated by computers at the Jet Propulsion Laboratory in Pasadena, California. The JPL computers were programmed with information from thousands of satellite photos like the one at far left, which shows volcanoes of Tharsis Montes from 4,933 miles overhead. The resulting database can be tapped for oblique views from any point above the surface; the new perspective, at near left, shows the volcanoes from an altitude of twenty miles—only a few miles above the summit of Arsia Mons *(foreground)*.

impression was reinforced by the twin flybys of *Mariner 6* and *Mariner 7* in July and August of 1969.

Because of Mars's periodic proximity to Earth, the summer of 1971 was a particularly good time to dispatch additional reconnaissance probes. Launched during this window, NASA's *Mariner 9* arrived at Mars in mid-November—just in time for an outbreak of planet-girdling windstorms. From telescopic observations, astronomers knew that the thin Martian atmosphere could occasionally whip itself into a fury, stirring up billows of ruddy dust that utterly obscured the surface. Explanations proposed for the phenomenon were little better than speculation, but whatever the cause, *Mariner 9* had to contend with the most severe outbreak since 1924.

Eventually, however, as months passed and the dust settled, amazing sights came into view. First to appear were the craters of ancient volcanoes—immense circular mounds that towered more than sixteen miles above the mean height of the Martian landscape. Elsewhere, thousand-mile-long canyon complexes cleaved the surface, and vast tracts of undulating sand dunes filled frame after frame from *Mariner 9*'s cameras. Icecaps at the poles appeared to be delicately layered, suggesting eras of intense cold alternating with periods of warmer weather. But expectant scientists were most surprised by the discovery of long, sinuous valleys that looked exactly like desiccated riverbeds on Earth.

The discovery of ancient Martian waterways only heightened the anticipation for the next round of robotic visitors, the double-barreled Viking missions of 1976. Viking consisted of two identical spacecraft, each made up of an orbiter and a lander. On July 20, as *Viking 1*'s orbiter photographed the planet and analyzed its atmosphere, the lander settled gently onto a smooth region of Mars named Chryse Planitia, the Plains of Gold. The lander's camera returned photographs of sand and rocks, gently rolling horizons, and a bright sky—scenery reassuringly similar to that found in the deserts of the American Southwest.

But Viking's prime purpose was not photographic. Engineers had equipped the craft with a small digging device similar to a backhoe, enabling it to dig shallow trenches in the Martian soil. A scoop collected samples for delivery to on-board biological experiments that, scientists hoped, might discover signs of life on the planet. Disappointingly, the tests detected nothing alive. Nor did they turn up any trace of the chemical compounds that even long-extinct microscopic organisms would have left behind. *Viking 2,* which landed in the Utopia Planitia on September 3, also came up empty-handed. As a worthy goal for human exploration, NASA found Mars's attractions markedly diminished.

KEEPERS OF THE FLAME

Nevertheless, the idea of voyaging to Mars had built up a following among enthusiasts outside the government. During the early 1970s, as the Viking spacecraft were being designed and built, a new grass-roots space-advocacy group called the National Space Institute (now the National Space Society) was founded. Wernher von Braun, who by then had retired from NASA, became president of the group.

Almost a decade later, a loose confederation of graduate students at the University of Colorado, collaborating with an informal network of space-travel aficionados from around the country, started working up a detailed outline for travel to the fourth planet. In 1981, they took the bold step of organizing a conference entitled "The Case for Mars." To be held the following year, the colloquium would address the feasibility of human visits to the Red Planet. Word of the event spread quickly throughout the community of space scientists. "Somehow they heard about the conference and volunteered," recalls co-organizer Christopher McKay. "It really was a Mars Underground."

The meeting convened at the end of April 1982 in Boulder, Colorado. In three days of spirited discussion and debate, scientists and engineers made a strong argument that heading Marsward could be—and should be—the next major goal of the U.S. space program. In addition to its main purpose of keeping the spirit of space pioneering alive, the conference served as a forum for assessing the state of the technologies that would be needed to reach Mars and for gathering new ideas to make such a mission practical sooner rather than later. Aerospace engineer James French, for example, described how a procedure called aerobraking could cut fuel requirements for the trip virtually in half: A spacecraft would slow to Mars-orbit speed by using friction generated from skimming through the thin Martian atmosphere rather than by braking with its engines. Then a lander would descend to the planet's surface.

Another proposal envisioned sending crews to one of Mars's two small moons, Phobos and Deimos. Called the Ph-D mission, it was considered by

STOPPING THE MARS EXPRESS

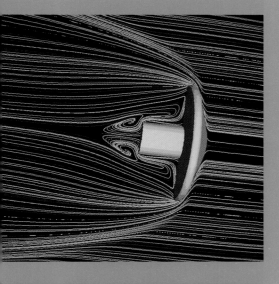

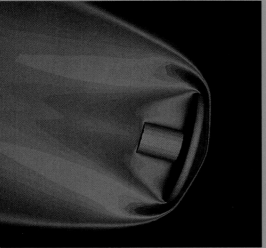

One of the most critical phases of a piloted mission to Mars begins as the spacecraft nears its destination: Moving at a velocity of more than 12,000 miles per hour, the ship must slow to just 1,700 miles per hour. This speed will put it into an orbit with a low point about 300 miles above the surface, from which a lander can safely descend.

The traditional means of braking a spacecraft is to fire retrorocket motors that counter the motion of the vehicle. But the rocket fuel required for braking a Mars-bound ship would make up 40 percent of the vehicle's launch weight, sharply limiting the capacity for food, water, oxygen, and scientific and life-support equipment.

A technique known as aerocapture provides a weight-saving alternative. Instead of braking rockets, the bow of the ship will bear a broad, saucer-shaped structure called an aerobrake. As the craft plows into Mars's atmosphere, it will decelerate rapidly because of the drag produced by the blunt aerobrake; at the same time, the aerobrake will also provide lift that keeps the vehicle airborne. A guidance computer will monitor the trajectory, changing the ship's orientation to maintain a constant level of drag. As soon as the spacecraft has slowed to the desired speed, the computer will roll the aerobrake into an attitude that provides maximum lift, causing the ship to pop back out of the atmosphere into an elliptical orbit. Later, a brief firing of the rocket's engine will shift the craft into an orbit just above the atmosphere. The spacecraft's angle of entry into the Martian atmosphere must be almost perfect. If it is too steep, the frictional heating will burn through the aerobrake and the ship's thermal protection, incinerating the spacecraft. If the angle is too shallow, the vehicle will skip off the top of the atmosphere and swing into a highly elliptical orbit, too far from the planet to be useful. In addition to providing the optimum combination of drag and lift, the aerobrake must protect the payload behind it from the blistering heat of its atmospheric passage.

Because they cannot test aerobrake designs in the Martian atmosphere, engineers use supercomputer models to study the effects of different variables. The picture at top left simulates the flow of nitrogen (the most common Martian atmospheric gas after carbon dioxide) around an aerobrake-equipped spacecraft plunging into the Martian atmosphere at 20,000 miles per hour. The bottom picture shows heat buildup around the same vehicle, ranging from searing temperatures of 21,000 degrees Fahrenheit *(dark red)* to a relatively cool 2,000 degrees *(dark blue)*. Still unknown is whether human passengers, after a long trip in weightlessness, could withstand deceleration forces more than five times as strong as Earth's gravity.

NASA's *Mars Observer* probe, pictured here surveying the Red Planet from orbit, is designed primarily to monitor the weather and study the composition of the Martian surface and atmosphere. Planned for the early 1990s, the probe will also carry a camera—capable of detecting objects smaller than five feet across—that may be used to examine potential landing sites for a sample-return mission.

some to be an attractive precursor—or even alternative—to landing on the planet itself. From a laboratory established on one of the satellites, scientists would dispatch rovers to fetch rocks and soil samples from Mars for analysis. Such a lab could have more and better equipment and could be established sooner and at less expense on Phobos or Deimos than on Mars, if only because fuel conserved by landing in the low gravity of either moon could be replaced in part with a greater payload of instruments, including such bulky items as electron microscopes. Furthermore, a base on Phobos or Deimos would permit planetary scientists to study those mysterious moons and perhaps divine their origin.

During the next three years, so much new thinking on Mars emerged that a second conference was held, in July 1984. Conference seminars delved deeply into such topics as turning carbon dioxide from the planet's atmosphere into rocket fuel made up of the gas's components—carbon monoxide and oxygen—which together burn vigorously. Details of spacecraft design were examined, as were the complex logistics for such a mission, and the human social and psychological factors that must be taken into account in the selection of crew members.

By the time of a third "Case for Mars" gathering in July 1987, talk of going to the planet was audible at the highest levels of government policymaking. In August, the staff of the newly established NASA Office of Exploration began to define how best to "expand human presence and activity beyond Earth orbit into the solar system." The division's first report, published in December 1988 and titled "Beyond Earth's Boundaries," presented a quartet of case studies, or "pathways," involving a broad range of capabilities, objectives, and timetables. Notably, three of the four scenarios called for sending crews to the Martian system. Earlier in the year, the Planetary Society's "Mars Declaration," endorsed by hundreds of prominent national figures, had urged spacefaring nations to cooperate on an exploration program culminating in piloted missions to Mars.

Despite the blizzard of words about Mars, there was little action. America's only nod toward the planet was to schedule a reconnaissance by an orbiting probe, closely resembling a weather satellite. Named *Mars Observer*, the probe is to be placed in Earth orbit by the space shuttle in 1992. From there, it will embark on a transit to Mars. The probe will enter an orbit that passes over the polar regions and conduct investigations for at least a full Martian year (687 days), its camera and laser altimeter mapping the planet surface as spectrometers and other sensors record seasonal weather changes and analyze the atmosphere.

JINXED?

During the decade and longer of American ambivalence toward piloted Mars flight, the Soviet Union gained a clear lead in preparing for interplanetary journeys. But the Soviets had also had their full share of problems. In 1962, they launched *Mars 1*, a flyby probe whose radios were somehow silenced en route to the planet. Nine years later, *Mars 2* and *Mars 3* were sent off. Upon reaching its destination, *Mars 2* dispatched a lander that probably slammed into the planet without slowing down. *Mars 3* reached the surface intact on December 2, 1971, but its transmissions ceased mysteriously only 110 seconds after touchdown. Four Soviet spacecraft sent to Mars two years later fared just as poorly, with only one orbiter working as planned. The entire Soviet Mars effort seemed jinxed, and the USSR would send no other craft to the Red Planet for the next fifteen years.

In the field of human endurance missions, however, the Soviets had been making significant strides with their space stations Salyut and Mir. Successive teams of cosmonauts occupied these outposts for ever-longer intervals. The durations of their ordeals in weightlessness—up to a year, in one instance—gave the Soviets the lead in space medicine, and high-ranking space officials, buoyed by these achievements, began to speak with confidence of sending cosmonauts to Mars.

To advance that cause, the Soviets scheduled a pair of unpiloted missions to inspect the planet's two satellites at close range. With picture-perfect launches in July 1988, *Phobos 1* and *Phobos 2* rocketed into space for the 200-day trip. However, the first of the probes was lost a few weeks later when ground controllers sent the craft a faulty command, inadvertently turning off the system that kept it oriented toward Earth. Radio contact was broken, and the probe began to rotate wildly. The other spacecraft managed to send back spectacular pictures of Phobos before radio transmissions faltered, then stopped altogether.

The setbacks seem not to have deterred the Soviets. Next on the timetable is a pair of orbiters to be launched toward Mars itself in 1995. Each will dispatch to the surface several small probes and a slightly heavier-than-air balloon instrumented for both atmospheric and surface reconnaissance. Two years later, the Soviets hope to send to Phobos a lander that will scoop up sand and rocks and then return them to Earth. Unfortunately, however, the prospects for one project, a piloted flight shortly after the turn of the century, have dimmed. The Soviet government has balked at the immense costs of such a mission.

A MATTER OF ENGINEERING

Before representatives of the human race—Soviet, American, or any other nationality—can leave footprints on the face of Mars, not only must funds be forthcoming but also a host of technological challenges, many with economic implications, must be surmounted. American and Soviet Mars-mission concepts, for example, both envision assembling huge interplanetary spacecraft, weighing at least one million pounds, by lifting components into low Earth orbit. Fuel and supplies for the trip, amounting to a large percentage of the vessel's weight, would also be carried aloft and loaded onto the Mars-bound ships. At space-shuttle rates of more than $2,000 per pound, the cost of readying a crew ship for Mars would be prohibitive. Clearly, a less expensive system must be developed, perhaps using unpiloted, expendable rockets or a crewless version of the shuttle that could carry more cargo per flight.

Great savings could also be achieved simply by reducing the amount of provisions needed for the trip. Each day, for example, an average-size person takes in about one and a half pounds of food, six pounds of water, and a little more than two pounds of oxygen. Using these values, a crew of eight on a two-year trip to Mars and back would use up approximately fifteen tons of water alone—without ever taking a shower. To alleviate the problem of supplying this much water for a long space mission, as well as to save the fuel expended in transporting supplies, the ship might well be built as a Controlled Ecological Life-Support System (CELSS), which recycles or regenerates consumables. NASA began tests on a CELSS prototype at the Kennedy Space Center in 1974, but by then Soviet scientists had been working on the problem for nearly two decades.

In 1972, the Soviets initiated their most recent experiment, called BIOS 3, near the Siberian city of Krasnoyarsk. At a complex about the size of the U.S. Skylab space station, crews of two or three lived for five months at a stretch, cultivating their own food from seed grown within the complex. Among the crops harvested were wheat, peas, potatoes, carrots, and tomatoes, variety

A Two-Stage Mission

A prime concern in sending humans to Mars and back is to keep the trip short, reducing the crew's exposure to the hazards of space travel. The laws of orbital mechanics dictate the most economical launch windows. An outbound or returning spacecraft must plan its departure and trajectory so that it will cross the target planet's orbit around the Sun when the planet is there to meet it. The launch windows, in turn, determine how long the crew stays on Mars.

As it happens, the shortest round trip requires the most fuel. It uses a flight path known as an opposition class trajectory because the round trip takes place within a twenty-six-month period between Earth-Mars oppositions—an alignment that puts Mars and the Sun 180 degrees opposite each other in Earth's sky. The most fuel-efficient mission flies a conjunction class trajectory (so called because at some point in the mission, usually during stopover, the planets move into conjunction, with Mars and the Sun on the same side in Earth's sky. This type of mission makes more efficient use of the Sun's gravitational pull, creating a kind of downhill run that lets the craft do the most coasting. But the elapsed time is three years, as much as half of it spent on the planet. The best choice may be a blend. With a "split-sprint mission" *(right and pages 90-93)*—one of several strategies NASA is studying—an unpiloted cargo ship takes a conjunction class path to Mars, where it is joined by a crew ship flying a faster, opposition class mission. The crew's trip time, including twenty days on Mars, is a year and a half.

Over several months, heavy-lift launchers carry pieces of the unpiloted Mars cargo ship into low Earth orbit for assembly by space-station workers. In addition to the vehicle itself *(top arrow)*, these include a Mars lander for descent to the surface, a rocket motor that will be used to inject the ship into trans-Mars orbit *(middle arrow)*, and a fueled propulsion system for the crew's return journey *(bottom arrow)*.

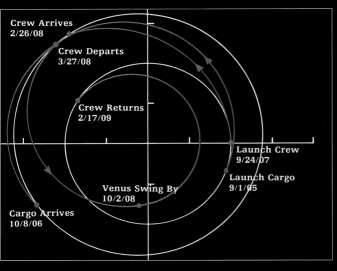

Crew Arrives
2/26/08

Crew Departs
3/27/08

Crew Returns
2/17/09

Launch Crew
9/24/07

Launch Cargo
9/1/05

Venus Swing By
10/2/08

Cargo Arrives
10/8/06

For a split-sprint mission starting in 2005, an unpiloted cargo ship would take thirteen months to reach Mars. The crew would start two years later and take five months. In this version, above, the crew ship swings by Venus on the return voyage, using the planet's gravity to bend the ship's trajectory so it catches up more rapidly with Earth.

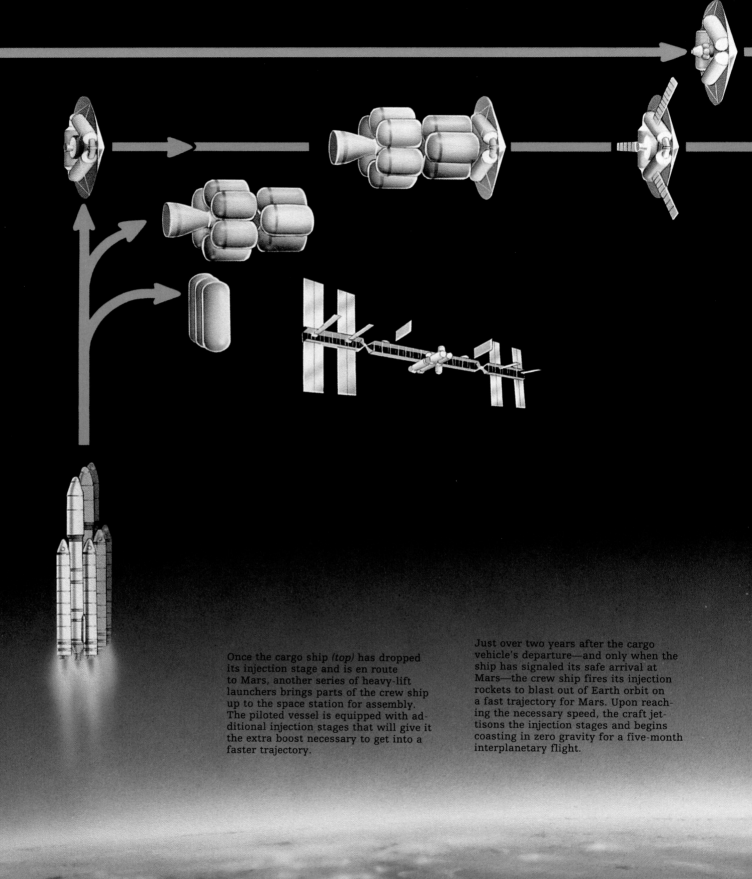

Once the cargo ship *(top)* has dropped its injection stage and is en route to Mars, another series of heavy-lift launchers brings parts of the crew ship up to the space station for assembly. The piloted vessel is equipped with additional injection stages that will give it the extra boost necessary to get into a faster trajectory.

Just over two years after the cargo vehicle's departure—and only when the ship has signaled its safe arrival at Mars—the crew ship fires its injection rockets to blast out of Earth orbit on a fast trajectory for Mars. Upon reaching the necessary speed, the craft jettisons the injection stages and begins coasting in zero gravity for a five-month interplanetary flight.

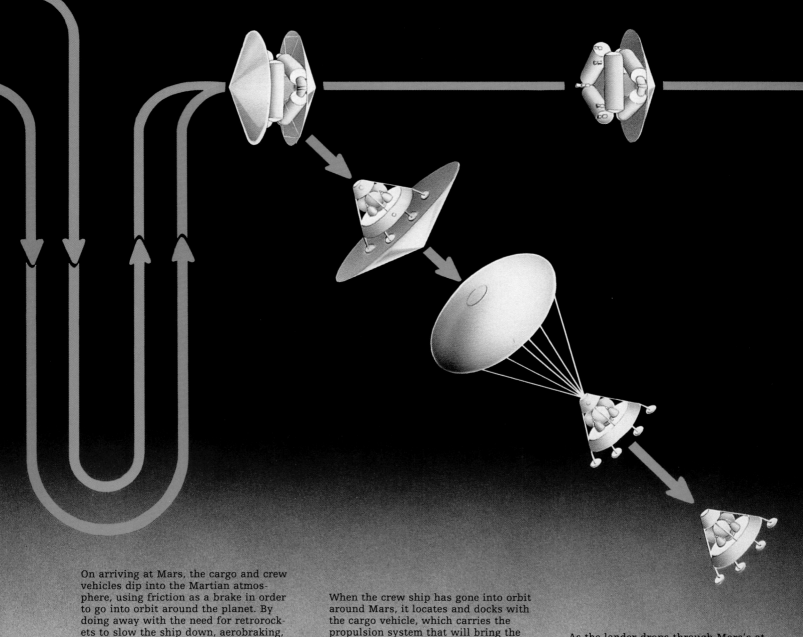

On arriving at Mars, the cargo and crew vehicles dip into the Martian atmosphere, using friction as a brake in order to go into orbit around the planet. By doing away with the need for retrorockets to slow the ship down, aerobraking, as the technique is known, saves fuel. But it requires precision: Too steep an angle of entry into the atmosphere will turn the ship into a fireball; too shallow and the ship will skip like a flat stone on water, sailing into interplanetary space with no hope of rescue.

When the crew ship has gone into orbit around Mars, it locates and docks with the cargo vehicle, which carries the propulsion system that will bring the crew back to Earth as well as the small Mars lander for descending to the planet's surface. Four crew members detach the lander and prepare it for descent; the others transfer the propulsion system to the mother ship.

As the lander drops through Mars's atmosphere, aerobraking cuts most of its speed. Parachutes slow the craft further but are relatively ineffective in the thin air—about the same density as air at 120,000 feet on Earth. A mile or so from the surface, the crew jettisons the parachutes and fires a retrorocket that slows the craft, giving the astronauts time to choose a landing site.

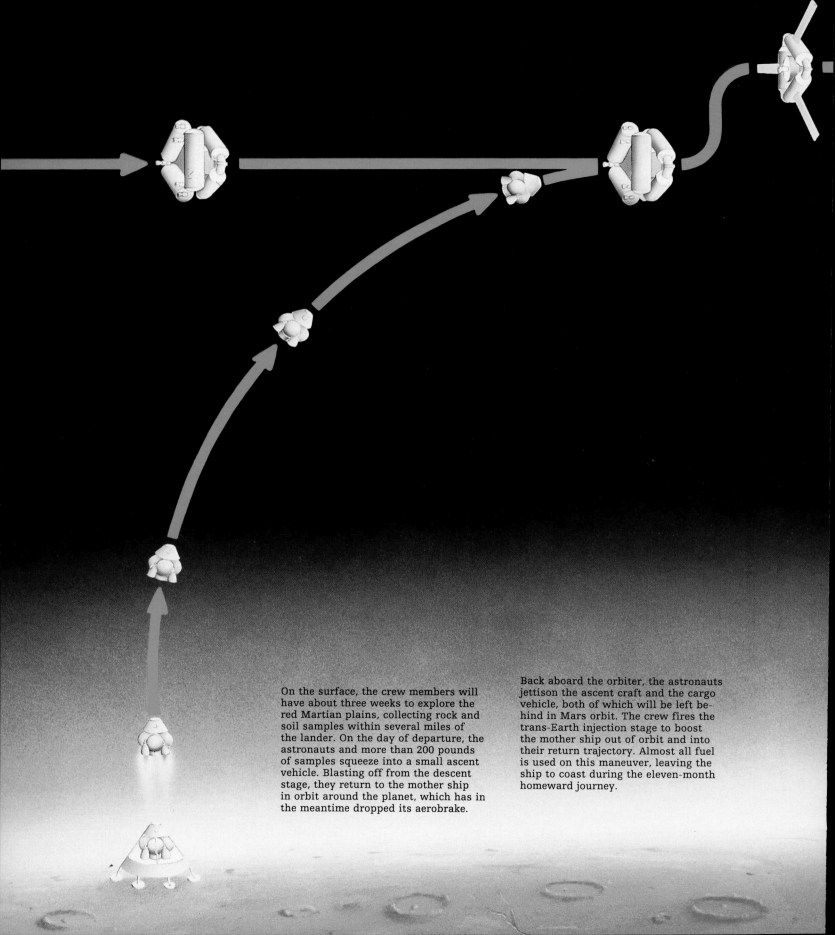

On the surface, the crew members will have about three weeks to explore the red Martian plains, collecting rock and soil samples within several miles of the lander. On the day of departure, the astronauts and more than 200 pounds of samples squeeze into a small ascent vehicle. Blasting off from the descent stage, they return to the mother ship in orbit around the planet, which has in the meantime dropped its aerobrake.

Back aboard the orbiter, the astronauts jettison the ascent craft and the cargo vehicle, both of which will be left behind in Mars orbit. The crew fires the trans-Earth injection stage to boost the mother ship out of orbit and into their return trajectory. Almost all fuel is used on this maneuver, leaving the ship to coast during the eleven-month homeward journey.

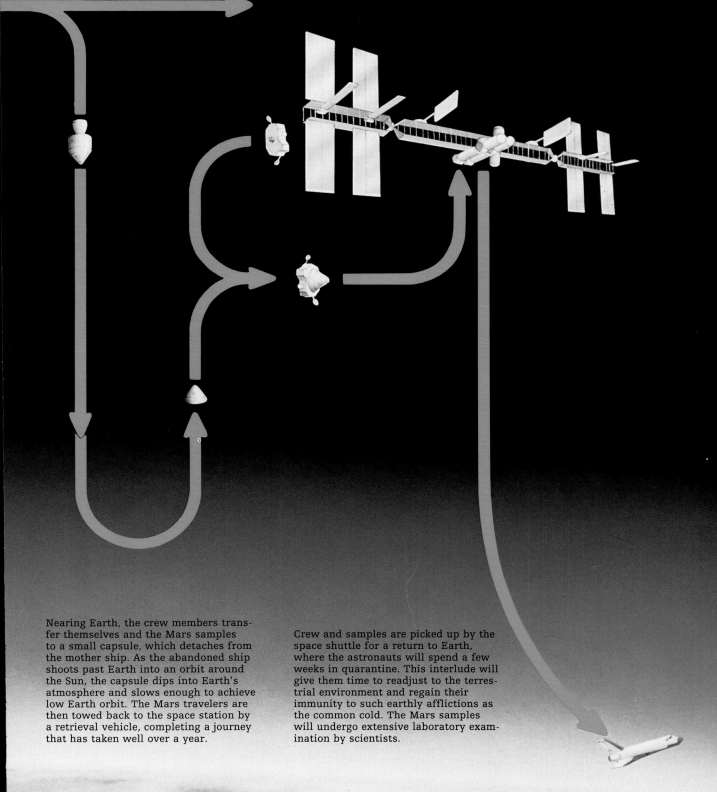

Nearing Earth, the crew members transfer themselves and the Mars samples to a small capsule, which detaches from the mother ship. As the abandoned ship shoots past Earth into an orbit around the Sun, the capsule dips into Earth's atmosphere and slows enough to achieve low Earth orbit. The Mars travelers are then towed back to the space station by a retrieval vehicle, completing a journey that has taken well over a year.

Crew and samples are picked up by the space shuttle for a return to Earth, where the astronauts will spend a few weeks in quarantine. This interlude will give them time to readjust to the terrestrial environment and regain their immunity to such earthly afflictions as the common cold. The Mars samples will undergo extensive laboratory examination by scientists.

enough to supply a balanced diet for BIOS 3 inhabitants. The growing plants converted carbon dioxide exhaled by the crew into oxygen for breathing, and they discharged water vapor that was condensed for drinking. Sewage yielded fertilizer and water for the crops. By the time the experiment ended in 1986, BIOS 3 had achieved 95 percent autonomy; only 5 percent of the crew's needs had to be supplied from outside the system.

An American CELSS project is even more ambitious. Called Biosphere II, it is funded by venture capitalists hoping to cash in on sales of environment-preserving inventions derived from the experiment. Covering only three acres in Arizona, Biosphere II is intended to be entirely self-sustaining. Eight people will dwell there continuously for two years beginning in 1990, living off 150 kinds of crops, many species of domesticated animals, including Vietnamese pygmy goats, and the bounty of a "wilderness" where game can be harvested.

Ecological ingenuity is not the only way to attack the payload problem. In 1986, some physics students at the University of Texas, intrigued by the prospect of interplanetary travel, were pondering a mission to Mars

Weakened by eight months of weightlessness aboard the *Salyut 7* space station in 1984, three cosmonauts rest in special lounge chairs outside the spacecraft that has returned them to Earth. Their mission, the same duration as a one-way trip to Mars, showed that interplanetary explorers might be incapacitated for days after reaching their destination.

when they came to a fuel-saving realization: The crew and its "luggage" need not fly together. For example, an unoccupied cargo vessel could follow a slow, fuel-efficient Hohmann trajectory to Mars, where it would fire retrorockets or use aerobraking to slip into a parking orbit after a trip of, in this case, nine months. The payload would consist of landing vehicles and other equipment needed specifically to explore Mars, plus all the supplies for the return trip.

Only upon safe arrival of the cargo ship in orbit around Mars would the crew be dispatched from Earth. Using a somewhat faster path to minimize transit time, they would rendezvous with their luggage after a little more than seven months en route. (The return voyage to Earth, begun a few weeks later, could take as little as five and a half months.) This flight plan for Mars—the so-called split-sprint trajectory—has been investigated and refined by engineer John Niehoff of Space Applications International Corporation. According to Niehoff, a pair of aerobraking spacecraft might weigh as little as 300,000 pounds each fully fueled and provisioned. One version of a split-sprint mission, designed by the aerospace firm of Martin Marietta Astronautics Group, is depicted on pages 89-93.

IN GOOD HEALTH

Aside from its fuel efficiency, a split-sprint mission may be necessary to preserve the health of the astronauts. The Soviet cosmonauts who circled the Earth for a year aboard Mir lost calcium from their bones at the rate of .5 percent each month. A longer journey to Mars and back could bring that figure uncomfortably close to the 25-percent calcium loss that makes the human skeleton brittle and easily fractured. The crew will also be subjected to harmful cosmic rays—high-speed particles originating deep in space. And there is the possibility of a dangerous burst of radiation from the Sun. Not only is solar radiation more intense outside the Earth's protective magnetosphere, but also sudden eruptions on the Sun's surface can spew lethal doses of high-speed protons, x-rays, and other forms of radiation outward into the Solar System. To protect astronauts from such events, advances in radiation shielding will be necessary. Lead, the material that most often fills this role, is far too heavy for a spacecraft; shielding materials of lower density must be developed.

Just as worrisome as the potential hazards of traveling to Mars and as challenging as the technical obstacles to be overcome is the extremely high price of such missions. In the summer of 1989, when President Bush advocated that the United States land on Mars, the cost of the initial step alone—a huge space station situated in Earth orbit—was estimated at $20 billion. A total of perhaps $400 billion would be needed to set up a small Martian colony. Sums of this magnitude argue for a cooperative effort by an international team, perhaps with Europe, the United States, Japan, and the Soviet Union as players. Even with faultless coordination, few technical setbacks, and a sufficiency of funding, only the most confirmed and unabashed optimists envision a permanent human presence on the Red Planet much sooner than fifteen years after the start of such a project....

OUTPOST

... July 2021: As four cargo vessels orbit Mars, filled with equipment ranging from knives and forks to engines built for a fuel mix of carbon monoxide and oxygen, a pair of space-going vessels from Earth, each carrying a crew of five men and women, aerobrake into a parking orbit above the supply ships. The events that are about to unfold culminate two decades or so of the most thorough reconnaissance imaginable of the planet and its moons, Phobos and Deimos.

On the third orbit, the spacefarers rendezvous with the supply ships and take on enough hydrogen propellant, water, oxygen, and other commodities to see them safely back to Earth should something go awry during the next step. Then they send a radio signal that awakens the supply ships' computers. Beginning a completely automated sequence, the computers fire rockets briefly, slowing the ships and kicking them out of orbit to begin the descent. Their destinations are two landing sites, each less than a mile from a base. Once briefly occupied but now vacant, they had been established several years earlier near the mouth of a long-dry waterway named Mangala Vallis. More a gorge than a valley, Mangala Vallis had been selected as a research site because its steeply eroded walls revealed eons of geologic history, including a deposit of lava four miles thick that covers much of the northern hemisphere under the planet's ancient, asteroid-battered crust. Computers aboard the supply ships report safe landings by all; the mission can proceed as planned.

Descent modules carry all ten astronauts to the Martian surface, depositing them a short walk from the Mangala Vallis installation, which will serve as the nucleus of a larger base that the astronauts will assemble from prefabricated parts aboard one of the supply ships. First things first, however. The colonists unload a rover and a trailer to haul food, oxygen, and water from the supply ships.

The explorers will not depend on this source for long. Within just a few days, for example, a water-extraction system is up and running. Powered by sunlight during the day and batteries at night, the machinery compresses the thin Martian atmosphere, collects the little water it contains, then breaks down its plentiful carbon-dioxide molecules to yield carbon monoxide and oxygen. Although considered only second-rate among rocket propellants, this combination of gases burns readily enough to provide both heat at night and fuel for the rovers. Some of the oxygen is siphoned off for use in the crew's living quarters and greenhouses; soon, the base will become an even more efficient and elaborate CELSS than the spaceships that brought the explorers to Mars.

Included in the group are a doctor, a botanist, two geologists, and a pair of systems engineers. Chosen as much for their long-term compatibility as for their professional specialties, they represent the United States, the Soviet Union, Japan, and France. Though these individuals are not the first to set foot on Mars, they might be considered the planet's original colonists. Their predecessors were on the surface for only six months or so before returning to Earth and had never become self-sufficient. The latest arrivals plan to remain four years.

The Martian environment soon lives up to its reputation for harshness. Twenty-three days after their arrival, the sky seems oddly hazy overhead, and more so to the southwest. By the next morning the base is engulfed in a choking cloud of dust. Raised by seventy-mile-per-hour winds hundreds of miles away, it has drifted over the colony and covered everything with a fine yellow powder. Fortunately, when the colonists emerge from shelter three weeks later, they find the all-important water-extraction system undamaged, and the work of assembling the prefabricated structures commences. By the end of the summer, they will have expanded the Mangala site into a permanent settlement—a nest of interconnected modules that have been partially buried to offer both insulation and protection from radiation. Once the settlement is on sound footing, the adventure of exploring the planet can begin.

A MATURE COLONY

Fifty years pass, and not only has the Mangala Vallis base remained continuously occupied, but its contingent of "Martians" now numbers more than 100 scientists and engineers. Remote outposts have been established where other scientists live and work—one, for example, in Mare Boreum, the great sea of sand dunes surrounding the north polar cap, and another at the base of a huge extinct volcano. Nuclear energy has become the source of power at all the Martian stations. Colonists at Mangala Vallis use it to fuse sand from the ancient waterway into sturdy glass plates that will be used in future construction at the base.

Communication with Earth is frequent, but not chatty. With transmissions taking between three and twenty minutes to span the interplanetary void, two-way conversation is trying, at best. However, one-way broadcasts of news and other programs are beamed continuously from Earth. This day, the Interplanetary Resource Consortium (IRC), formed two decades earlier to promote commercial mining on the Moon, reveals plans to set up a methane-processing plant on Titan, Saturn's large, murky satellite. The announcement had been expected; the IRC recently built a metal-extraction plant on one of the small asteroids that cross Earth's orbit.

Much more significant for Mars, says the chief scientist at Mangala Vallis as a footnote to the newscast, is the solid rumor of a detailed IRC proposal to terraform, or endow the entire planet with an atmosphere and climate hospitable to terrestrial life. Since the confirmation ten years ago that vast quantities of water and carbon dioxide had been captured inside the planet at birth instead of being freed—as they had been on Earth—to form an atmosphere, only inertia and a shortage of funds stood in the way of terraforming Mars. Now, those obstacles seem to have fallen.

The initial step in terraforming would be to release these two critical ingredients into the atmosphere. Carbon dioxide and water vapor are transparent to incoming sunlight, but they trap heat radiating upward from the surface. Because of this so-called greenhouse effect, Earth's surface is some sixty degrees warmer than it would be without an atmosphere. Mars's average surface temperature is expected to rise above the freezing point of water.

The plan is for mirrors to be placed in stationary orbit over the poles to focus the Sun's feeble rays on the icecaps, raising their temperature and turning them to cold wisps of vapor. As the greenhouse effect takes hold, Mars will become warmer—imperceptibly at first, but more rapidly as decades pass. After a hundred years or so, all the water ice and frozen carbon dioxide in the caps will have been driven into the atmosphere, increasing atmospheric pressure until it is about one-tenth that of Earth—great enough for water to exist in liquid form without instantly evaporating. Oceans on Mars are probably too much ever to hope for, but lakes, streams, and rainstorms are likely. By that time, solar radiation will no longer be a threat; a well-developed ionosphere will sit atop the thick blanket of air. Martians of the future will be able to venture out of their domes equipped with nothing more than scuba-style breathing apparatus.

In the meantime, botanists will have developed flora—perhaps merely bacteria at first, then algae and lichens—that can thrive in an atmosphere rich in carbon dioxide but devoid of oxygen. (Although terrestrial plants require carbon dioxide for photosynthesis, they die in the total absence of oxygen.) Over thousands of years, these bacteria, algae, and other varieties of plant life will convert much of the global blanket of carbon dioxide into oxygen and carbon. Warming steadily, Mars's permafrost will begin to melt, releasing vast additional quantities of water and carbon dioxide into the new atmosphere and strengthening the greenhouse effect. If, after a hundred millennia of such activity, terraforming succeeds, Martians—whose origins on Earth will have become as much myth as history—will throw away their scuba gear and dismantle their CELSS domes. They will vacation at lakeside retreats and enjoy their planet's cool, dry climate as if their ancestors had known no other.

When the first colonists step onto the rusty sands of Mars sometime in the next century, the challenges they face will be as vast as the desolate landscape before them. Temperatures plunging to − 220 degrees Fahrenheit will be routine, deadly radiation a fact of life, and immense dust storms a constant threat.

With Earth millions of miles away, the colonists' long-term survival will depend on their ability to conserve the essentials of life—water, food, and oxygen—while exploiting the planet's natural resources. In the colony illustrated on the following pages, these functions are regulated by a complex recycling process known as the controlled ecological life-support system (CELSS). The highly automated system will maintain a delicate symbiotic relationship between the plants, animals, and humans in the colony while extracting substances such as oxygen and hydrogen from the Martian environment. By squeezing such vital elements from the reluctant planet, colonists in the CELSS will eventually achieve complete independence from Earth.

CELSS must both provide the Mars dwellers with an Earth-like atmosphere within the habitat module and maintain a carbon-dioxide-rich environment in the greenhouse for the plants. The system must also supply each colonist with an average daily ration of food, clean water, and oxygen estimated at nine and a half pounds, and take back for recycling the same amount of material in the form of carbon dioxide and solid and liquid waste. For example, colonists will exploit the normal respiration of greenhouse plants, using them to "inhale" carbon dioxide given off as waste by the humans and "exhale" oxygen, which will be pumped into the living quarters. A separate structure will house a recycling module where vats of algae, bacteria, and yeast break down both plant and human waste to provide fresh water, nutrients for the greenhouse, and cellulose for manufacturing.

A COLONY BUILT FROM SPARE PARTS

A months-long journey through interplanetary space ends as astronauts in a landing craft drift down under parachutes to join the bustling new colony. In the stark red reaches of a Martian valley, early colonists have established the nucleus of the base by tipping over conical landing-craft shells for use as living quarters, medical facilities, and scientific laborato-

ries. A greenhouse clings to the tail section of one such habitat, which has had its rocket nozzles removed.

Out on the frigid Martian surface, an automated dragline and scoop cover the living areas with soil. The dirt will act as a shield against the solar and cosmic radiation that so easily penetrate the thin Mar-

At right, a compressor sucks in the carbon-dioxide-laden atmosphere for conversion into its constituents—oxygen and carbon monoxide. Stored in large tanks, the oxygen supplies the habitat module, while the carbon monoxide, in liquid form, is used for rocket fuel. Arrays of solar panels flank the colony, providing

A High-Tech Haven for Earth's Plants

Colonists nurture the carefully selected plants and aquatic animals in their greenhouse through a mixture of farming techniques. Among them are hydroponics, in which plants rooted in an inert substance (possibly Martian soil) receive liquid nourishment; aeroponics, in which plants are suspended on a perforated frame

tional, soil-based agriculture, using soil from Mars.

The settlers are vegetarians, living on wheat, potatoes, soybeans, strawberries, and other high-protein, high-calorie foods. In one corner of the structure, an experimental aquaculture tank hosts a small colony of shrimp and crayfish, which not only provide food but also eat dead plants and produce wastes for fertilizer.

The transparent, plastic double walls of the greenhouse, highly resistant to ultraviolet light, enclose a mix rich in carbon dioxide and water at less than one-half of Earth's sea-level pressure, an optimal environment for the crops. A computer near the air lock regulates pumps and filters to control the flow of water, nutrients, waste, and gases.

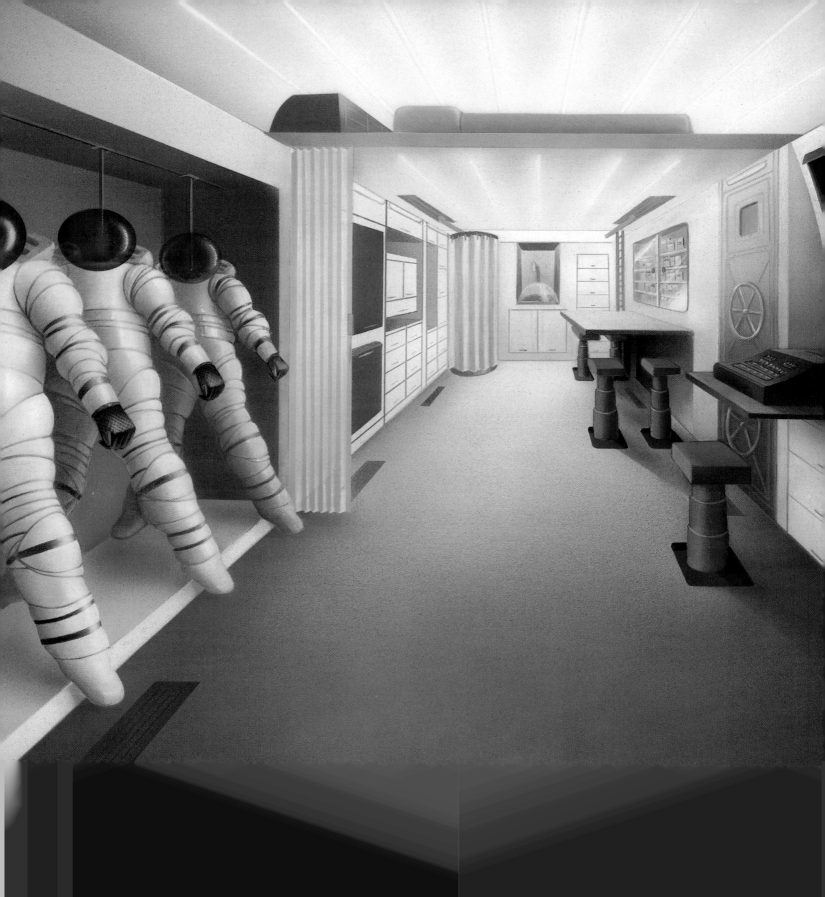

made from compressed plant cellulose, are designed to fold out of the walls and floor for maximum efficiency. With space at a premium, the bedroom is limited to a loft near the ceiling. Fluorescent lights complement a fiberoptic system that draws in certain beneficial wavelengths of natural light while allowing harmful ultraviolet rays to be removed.

In the foreground, an assortment of spacesuits, some for venturing outside and others for working in the partially protected greenhouse, stand ready next to a folding divider. A computer center dominates the area; the machines supervise the entire CELSS and monitor the colony's activities for any problems that might jeopardize the success of the operation.

The three stars that are the Sun's closest neighbors *(circled)*—Alpha Centauri A and B and Proxima Centauri—blaze in a corner of the southern sky. Near as they are, spacecraft powered by conventional chemical rockets would nevertheless have to journey for millennia to cross that 26-trillion-mile interstellar void.

rom deep within the single huge exhaust cone, a blinding radiance flares, only to vanish again as, with an almost imperceptible shudder, the massive starship eases out of Earth orbit on the first leg of its journey through interstellar space. During the next several days, the command crew puts the craft through an intricate series of maneuvers accompanied by more brief engine firings: first a close swing around the Sun to pick up a free boost from its gravitational pull, then a quick dodge up and over the asteroid belt, and finally two more gravity assists, at Jupiter and Saturn. By week's end, the ship's speed has climbed to well over a million miles per hour.

As the ship makes its roundabout exit from the Sun's domain to save precious fuel for the long voyage ahead, signs of humanity's spacefaring achievements are passed like milestones—a diadem of slowly spinning space-wheel colonies in orbit around Earth; satellites keeping watch over a mammoth terraforming project that will eventually give Venus a breathable atmosphere; mass drivers stationed near the asteroid-belt mines, hurling a steady stream of mineral payloads across millions of miles to supply industrial complexes on Mars and on Saturn's largest moon, Titan.

So far, the trip is little different from those of the first starships, the great interstellar arks of a hundred years ago—still only halfway to their destinations. But out beyond Neptune, when the captain gives the order for sustained thrust, this revolutionary craft surges forward on a wave of energy that will accomplish what was long thought impossible. Driven by the most sophisticated of the new breed of matter-antimatter engines, the ship will accelerate continuously for years, reach more than two-thirds the speed of light, and arrive at its goal—the planet-rich Arcturus system, more than thirty-six light-years from Earth—within the lifetime of those now watching their home star grow smaller by the hour.

Twenty years out, a historic milestone is recorded in the logbook: The thousand passengers on this great vessel have outdistanced the very first starship and reached deeper into space than any other humans. But the travelers are aware of an irony in this: Chances are that, just as they have caught up with those who started before them, future voyagers riding a still unguessable technological breakthrough will overtake them and sweep ahead, leading the human species into the ebony immensities of the universe and whatever it holds. . . .

THE STUFF OF DREAMS

. . . Spacefarers of the distant future will most likely be amused by twentieth-century speculations about interstellar travel. Much of the conjecture is pure science fiction, bearing little resemblance to possibilities grounded in current understanding of physics and the workings of the cosmos. Yet a surprising amount of detailed, scientifically respectable work has been done on the subject. One group has even drawn up blueprints for the first interstellar vessel, taking full account of fuel-to-payload weight ratios, exhaust velocities, and a host of other specifications. Schemes range from innovative adaptations of present technology to more ambitious concepts based on educated guesses about what will become feasible in the next hundred years or so. But so strange and unpredictable has the progress of science been that the most imaginative and seemingly unlikely approaches may prove to be closest to the mark. Science fiction may, in fact, hold the answers.

The crux of the starfaring problem is, of course, the virtually inconceivable distances of interstellar space. The nearest star system, Alpha Centauri, is some 26 trillion miles away, so far that its light, traveling at 186,000 miles per second, takes more than four years to reach Earth. Were it headed that way, the *Pioneer 10* Jupiter probe, now hurtling beyond the Solar System at 25,000 miles per hour, would take 110,000 years to cover the distance. Much of the work on interstellar travel has thus focused on developing radically new forms of propulsion that, by significantly boosting a craft's top speed, would shrink flight times to more manageable proportions.

But new spacecraft, no matter how vastly improved, will not bring even the nearest stars within easy reach. Starships traveling thousands of times faster than today's speediest rockets would still face voyages of decades or centuries to make the crossing. If humans are to be aboard, scientists will have to devise vessels and systems capable of sustaining life—or at the very least, the seeds of life—for hundreds of years. The technological hurdles of engineering new propulsion systems may turn out to be trivial compared with the difficulties of engineering people able to endure such journeys. Scientists have barely begun to explore the psychological and sociological effects of extremely long-term spaceflight—and as yet have no way to extrapolate their findings to generations born and raised on ships crossing the lonely oceans of interstellar space.

The true magnitude of the task of venturing starward became apparent only in the nineteenth century, with the first accurate measurements of the distance to Earth's closest stellar neighbors. Einstein's special theory of relativity, published in 1905, brought more discouraging news: The velocity of light was apparently an absolute cosmic speed limit that matter could never attain. In light of these findings, the prospects for visiting a neighboring sun were so daunting that most scientists were inclined to dismiss the subject as mere fantasy. Even those who were irresistibly drawn to the topic knew they risked being labeled cranks and, as a result, were somewhat circumspect about their intellectual investment in it. In 1917, Robert Goddard, who nine

American physicists Theodore Taylor *(far left)* and Freeman Dyson were the prime movers of Project Orion, a scheme for building spaceships propelled by nuclear explosions. In the late 1950s, Taylor asked Dyson to leave the Institute for Advanced Study in Princeton, New Jersey, and help him with the endeavor at the General Atomic Corporation in San Diego. Experiments with miniature vehicles and chemical explosives suggested that the technique might work, but a ban on atmospheric nuclear testing ended the project in 1965.

years later would build and launch the first liquid-fuel rocket, wrote a privately circulated paper entitled "The Ultimate Migration," which, he said, "is to be read only by an optimist." Goddard declared, "I truly believe that it will eventually be possible for human beings to be transported on expeditions to the nearest stars—leaving earth while they are in a kind of deep sleep, the way seeds 'sleep' over the winter, and waking up at their destinations as much as a thousand years later."

Not all starfaring enthusiasts were as reticent as Goddard. In his 1929 book *The World, the Flesh and the Devil,* the radical British scientist J. D. Bernal speculated that a hollowed-out asteroid might serve as a starship. As the provocative title suggests, Bernal, who was known to his peers as "the Sage" for his encyclopedic knowledge of subjects ranging from Chinese ceramics to molecular biology, was actually more interested in social dynamics and ethics than stellar voyaging. But science-fiction writers were quick to develop the theme, conjuring up a wide assortment of fantastic vehicles and missions. Some followed Bernal's and Goddard's lead, writing of huge multigenerational "ark" ships fanning out to colonize the galaxy. Others stretched the rules of physics to accommodate small, high-speed scout vessels on round-trip journeys of exploration, while still others sidestepped reality altogether with faster-than-light space cruisers that came and went as they pleased throughout the galaxy and even beyond.

In the years following World War II, science did not exactly catch up with science fiction, but certain developments—the advent of nuclear energy in particular—did spur a reexamination of the difficulties inherent in starflight. Although there are a number of ways of judging rocket performance in the context of an interstellar voyage, the critical base for comparison is engine efficiency, which reflects not only how well a particular design produces propulsive thrust but also the energy potential of the fuel. Chemical rockets, the kind now used by all spacefaring nations, employ a fuel such as kerosene or liquid hydrogen in combination with an oxidizer, usually liquid oxygen. The two substances are mixed together in a combustion chamber and ignited, producing hot gases that are expelled at high speed through the rocket's nozzle, thereby propelling the rocket itself in the opposite direction. The hotter the burn, the higher the pressure and thus the exhaust velocity, and the faster the rocket—at least in theory.

Rocket scientists measure efficiency in terms of specific impulse, or Isp (pronounced "I-S-P"), typically expressed as the number of seconds during which one pound of fuel will produce one pound of thrust. A high Isp thus translates into more thrust per pound of fuel. The best chemical rockets, such as the main engines of the space shuttle, have an Isp of about 450 seconds. These engines could not carry enough fuel to reach even one-hundredth of one percent of the speed of light, a rate at which it would take more than 42,000 years to reach the Centauri system. Researchers theorize that to be practical, a starship would need an Isp rating on the order of 1,000,000 seconds—clearly beyond the capabilities of the most efficient chemical engines to date.

With nuclear energy, however, that kind of performance seemed within reach. As was demonstrated in a series of tests in the 1960s, heating liquid hydrogen by passing it through the core of a nuclear reactor created extremely high exhaust velocities. But the best that the nuclear rocket could muster, in ground tests in the late 1960s, was an Isp of 1,000 seconds. While this more than doubled the performance of chemical engines and might be sufficient for trips to the Moon or Mars, it was still far short of what would be needed for interstellar flight.

A NEW USE FOR THE BOMB

At about the time this project was just getting under way in the mid-1950s, a small group of scientists at the Los Alamos Scientific Laboratory (now the Los Alamos National Laboratory), where the atomic bomb had been born, were conceiving a more unusual application of nuclear power. Physicist Freeman Dyson reflected both the group's inspiration and its purpose when he later wrote, "We have for the first time imagined a way to use the huge stockpiles of our bombs for better purpose than for murdering people. Our purpose, and our belief, is that bombs which killed and maimed at Hiroshima and Nagasaki shall one day open the skies to man."

The basic concept actually predated atomic weapons by half a century. In 1891, Hermann Ganswindt, a German inventor, proposed building a rocket that would be propelled by a series of controlled chemical explosions ignited in rapid sequence. The principle was soon to find practical application in the automobile's internal combustion engine, where the power to push pistons comes from carefully timed explosions of a gasoline-and-air mixture. In 1955, Los Alamos physicists Stanislaw Ulam and Cornelius Everett revived the idea as it related to rocketry, this time with a much more powerful nuclear kick.

The idea was enormously attractive, especially to scientists who were disturbed by the ethical implications of their work on nuclear weapons. The leaders of what came to be known as Project Orion were Theodore Taylor, a young designer of small, efficient atomic bombs, and physicist Dyson, whose duties with the British Bomber Command during World War II had left him with deep moral misgivings. For them, the prospect of beating swords into starships was irresistible.

The Orion scheme envisioned a ship 100 feet long and 34 feet in diameter, whose base consisted of a massive aluminum plate connected to the rest of the ship by a set of gas-filled, doughnut-shaped shock absorbers. Nuclear bombs would pop out of a hole in this so-called pusher plate like gumballs from a vending machine. Each second, several bombs would explode at a distance of about fifty feet from the pusher plate. The shock wave from the expanding nuclear fireballs would slam into the pusher plate, accelerating the vehicle. Because of the shock absorbers, the ride would be bumpy but endurable for the crew.

Early tests of the concept seemed promising. In 1959, a 250-pound prototype rocket called Hot Rod, propelled by chemical explosives, successfully reached an altitude of several hundred feet. But that was as high as Orion was ever to go. Taylor and Dyson, who had fully expected that one day they would actually travel to the planets in an Orion ship, had set "Saturn by 1970" as their goal. Those responsible for funding decisions had less lofty ambitions, however. The Department of Defense could conceive of no plausible military use for the Orion vehicle, and NASA, which had oversight for the department's nonmilitary projects, was more interested in proven chemical rockets for its Apollo program. Then, in 1963, the signing of the nuclear test ban treaty prohibiting nuclear detonations in the atmosphere sealed the project's fate. In 1965, after $10 million had been spent, Project Orion was canceled.

In the design for the Orion spaceship (shown here in a scale model), hundreds of atomic bombs would drop in sequence through a hole in the rear pusher plate and explode in rapid succession. The resulting shock waves would drive the pusher plate forward on shock absorbers that would cushion the crew cabin as the craft moved ahead.

FUSION'S BIGGER KICK

Dyson was undeterred. In 1968, while working at Princeton's Institute for Advanced Study, he proposed an Orion-style spaceship as large as a super-tanker, which would be powered by 300,000 one-megaton hydrogen bombs. The hydrogen bomb was the atomic bomb's more potent cousin, fueled not by fission but by fusion—the very power of the stars themselves. In a fusion reaction, atomic nuclei meld rather than split: Crushed together under conditions of enormous temperature and pressure, they release three to five times as much energy as is produced by a fission reaction. By firing off one hydrogen bomb every three seconds for ten days, Dyson's vessel would reach a velocity of three percent of the speed of light (psol) and could deliver a 45,000-ton payload, the equivalent of a small space colony, to Alpha Centauri in about 130 years.

This so-called pulse fusion rocket, crude as it seemed, became the inspiration for one of the most sophisticated and plausible starship designs yet proposed. New research in the 1970s, conducted principally by scientists at the Lawrence Livermore Laboratory in California, led to an important refinement in the concept of pulse fusion. Rather than depending on a large number of big, uncontrolled nuclear explosions, the new version of Orion relied on a very large number of relatively small, controlled detonations. Laboratory work suggested that fusion might be achieved by subjecting a tiny pellet of deuterium, a heavy form of hydrogen, and helium-3, a light form of helium, to an all-around bombardment by tightly focused beams of fast-moving electrons or, alternatively, by very powerful lasers. The energy imparted by the beams would cause the pellet to implode, touching off a fusion reaction. At this point, one of the advantages of fusion over fission would come into play. Rather than generating vast amounts of radioactive waste along with heat, fusion produces an expanding ball of superheated gas, called plasma, which could serve directly as a very potent propellant, guided out the exhaust nozzle by an intense electromagnetic field. There would be no need for a separate propellant, such as liquid hydrogen, as in some fission rockets, further adding to performance by substantially reducing the total weight of fuel needed for a given thrust. The Isp ratings of some design versions reached the theoretical goal for starflight of 1,000,000 seconds.

It was little wonder, then, that pulse fusion became the propulsion method of choice for the British Interplanetary Society's Project Daedalus, the first full-scale study of the feasibility of building and launching a starship before the end of the twenty-first century. Conducted between 1973 and 1978, the study produced detailed plans for an unpiloted interstellar probe employing only technologies—pulse fusion being one—with a reasonably good chance of being developed in the next hundred years. The Daedalus probe would weigh some 54,000 tons at launch, 50,000 tons of which would consist of 30 billion pellets of deuterium-helium fuel. Firing off the pellets at a rate of about 250 per second, Daedalus would accelerate for about four years, reaching a top speed of about 12 to 13 psol. The target was Barnard's star, a dim, red star

5.9 light-years away, selected both for its proximity and for the faint likelihood of its having a planetary system. About fifty years after setting out, the probe would deliver in the star's vicinity a 500-ton payload consisting of a fleet of smaller probes that would investigate any planets discovered earlier by the main probe's computerized detection system. Some six years after the encounter—long enough for radio messages to make it back to Earth—and more than half a century after launch, the grandchildren of the scientists who built Daedalus would finally receive whatever data had been collected.

The Daedalus study illuminated many of the problems and complexities that would face other starship designers. To begin with, obtaining the fuel would pose a major challenge. Rare on Earth, helium-3 would have to be mined from the lunar surface, a project that would add significantly to the expense of the interstellar mission itself. Despite its improvements over other designs, Daedalus still had a serious weight problem. The fuel-to-payload ratio remained high, meaning that Daedalus could not easily be adapted to provide for a crew; the added weight for life support would significantly reduce the maximum speed and lengthen the duration of the trip. Moreover, 50,000 tons of fuel represented only what was necessary to accelerate the vehicle. Slowing down at the other end of the journey, as a crew-carrying flight would demand, would greatly increase the fuel requirements, and if the objective were to send a scout party from Earth and then bring them back, the original fifty-year jaunt would turn into an odyssey lasting more than a hundred years.

Daedalus planners also had to take into account the fact that space is not actually empty. Each cubic inch of the interstellar void contains a few atoms of hydrogen that would slowly but surely erode the leading edges of the vehicle. Protection against the ravages of interstellar gas and random particles of dust would come from a berylli-

In a venture known as Project Daedalus, researchers from the British Interplanetary Society drew up plans for a ship that would be powered by the nuclear fusion of deuterium and helium-3. The fuel for the two-stage vehicle would be stored in two sets of spherical tanks—the largest measuring nearly 200 feet across—and would provide continuous acceleration for almost four years.

um shield, but because space debris grows thicker in the vicinity of a star, the shield itself would become degraded once the probe reached its destination. A collision at starship speed with even a one-gram mote of dust would produce an explosion equivalent to 150 tons of TNT. The Daedalus solution to the problem was to include a "dustbug" robot that would fly about 125 miles ahead of the main ship, dispensing a cloud of tiny dust particles that would collide with and vaporize anything in the way.

LOOKING FOR A FREE LUNCH

Even before the Daedalus project came to grips with the quantity of hydrogen in interstellar space, other thinkers had turned the problem on its head. One of the main obstacles to efficient interstellar propulsion is that much of the energy produced by a rocket engine is spent accelerating not payload but fuel yet to be burned. In a starship such as Daedalus, where acceleration was to continue for years, a great deal of fuel would be just along for the ride most of the time, wasting energy. According to Louis Friedman, director of the American-based Planetary Society, "It is not reasonable to consider interstellar flights with conventional rockets that have to carry fuel—any fuel." Instead, some theorists asked, why not simply scoop up interstellar hydrogen along the way and use it for fuel?

Known as the free-lunch approach, the concept was pioneered by Robert Bussard, a tall, beetle-browed American physicist and rocket engineer who had worked on the first prototypes of nuclear rocket engines. In a 1960 paper, Bussard proposed constructing a vehicle that would collect interstellar hydrogen and feed it to a fusion engine, where it would be superheated. Known as the interstellar ramjet, the Bussard design envisioned a relatively light vehicle of only 1,100 tons that would accelerate at a constant 1 g; that is, the sensation would be the same as that of a body resting on Earth—thus relieving the crew of the problems associated with weightlessness. Its distinguishing feature would not even be visible—a huge electromagnetic field, generated by the fusion engine and measuring more than a thousand miles in diameter, that would snag interstellar hydrogen and funnel it into the engine. An initial supply of fuel would be needed to accelerate the ship to about two percent of the speed of light, at which speed the ramjet would encounter enough hydrogen to begin functioning efficiently. At the destination, the magnetic field could be reversed so as to repel rather than collect hydrogen, and the electromagnetic scoop would become a kind of parachute for decelerating the ship. In practice, the ramjet was expected to achieve a top speed of about 15 psol, but in theory, with an unlimited supply of fuel, there was nothing to stop it from reaching light-speed itself.

Bussard's idea stirred a great deal of interest during the 1960s and 1970s, but all the attention eventually revealed what some scientists thought were fatal flaws in the original concept. For one thing, Bussard had not allowed for the drag created as a ship plowed through interstellar gas. And in any case, further calculations based on refined estimates about the density of interstellar gas suggested that the ramjet would require a much larger scoop, ranging anywhere from 600,000 miles to half a light-year in diameter—improbable dimensions.

Bussard moved on to other pursuits in the 1980s, among them a new type of fusion generator as a power plant for a trip to Mars, but the idea of free propulsion had fired the imaginations of more than a few within the community of interstellar visionaries. Inspiration also came from one of the most ancient of technologies: sailing.

As early as the 1960s, researchers had envisioned sails of ultrathin plastic or aluminum up to 150 miles across that would catch the pressure of sunlight much as a conventional sail catches the wind. However, as scientists quickly realized, this kind of propulsion—soon dubbed solar sailing—would work fine for interplanetary travel, but not past the boundaries of the Solar System. To sail beyond the range of sunlight and out to the stars called for a more ingenious approach.

In the late 1960s, Canadian engineer Phillip Norem dreamed up a giant array of space-based lasers that would fire a tight beam of energy at the sail of a starship and continue to push it on its way for as long as ten years of constant acceleration, at which point such a vessel might reach a speed as great as 30 psol. Another scheme, of more recent vintage, would use a similar type of space-based laser beam instead of a fusion reactor as the heating source for starship propellant.

As is the case with most other starship designs, one of the basic assumptions in the directed laser proposal was also its prime stumbling block. A laser mighty enough to drive a sailship to a star would have to generate at least a trillion watts, more than 300 times the capacity of today's largest power stations. And a scout mission that would decelerate at the target for intensive exploration and then return home would increase the power requirement to a whopping seven trillion watts, which might raise the costs back on Earth to prohibitive levels.

PROVISIONS FOR PASSENGERS

Even at their best, most designs still leave the nearby stars several decades away and more attractive targets—with a higher probability of having planets that could be colonized, for example—even more removed in time. Some researchers have thus concentrated on developing sophisticated probes that would do much if not all of the necessary scouting, perhaps preparing the way for human voyagers of a far-distant era. So-called nanomachines, for example (pages 117-121)—atomic-scale robots that would reconfigure and even reproduce themselves—are among the most intriguing concepts for interstellar exploration yet proposed.

But the urge to get humans to the stars soon remains strong, and the first step toward that goal will clearly have to involve starships capable of supporting the original crew, and perhaps several succeeding generations, for the rest of their lives. Spacecraft bound for the stars will have to be small worlds

A Self-Made Scope

The crucial first step in an expedition beyond the Solar System will be to pick a destination. Yet even with dramatic breakthroughs in the resolving power of telescopes, detecting hospitable landing sites from Earth will remain difficult. So far, scientists have only the vaguest clues to possible extrasolar planetary systems, the nearest lying 6.2 light-years away. A high-powered telescope on an asteroid near Jupiter, however, would notably improve celestial reconnaissance. From this vantage point, astronomers on Earth could scan neighboring solar systems for a suitable landfall.

Building a telescope in this lonely outpost may one day be accomplished by the futuristic techniques of nanotechnology, which envisions the advent of machines on a scale of nanometers, or billionths of a meter. Nanomachines, and the nanocomputers that accompany and control them, would manipulate matter atom by atom *(above)*. Embedded among the molecules of a tiny transmutable probe and fired off into deep space, the nanomachines—proliferating by the trillions and obeying built-in instructions—could convert an asteroid from a lump of frozen rock into a sleek and farseeing space telescope *(pages 118-121)*.

MID-JOURNEY TRANSFORMATIONS

Several key stages in the imagined journey of a space probe containing nanomachines are shown above. To overcome Earth's gravity and avoid atmospheric drag, the probe is launched from a planetary orbit approximately 150 miles high. The launch vehicle—an elec-tromagnetic cannon girded by twenty to fifty coils of wire—can fire only circular projectiles of high conductivity, so the probe begins its trek in the form of a half-inch-diameter silver ring. When an electric current is sent through the coils around the barrel, magnetic fields of intense repulsion arise inside the barrel and fling the probe from the muzzle at a speed of 22,000 miles an hour *(above)*.

Although the ring shape is ideal for launch, its small surface area keeps it from maximizing the Sun's en-

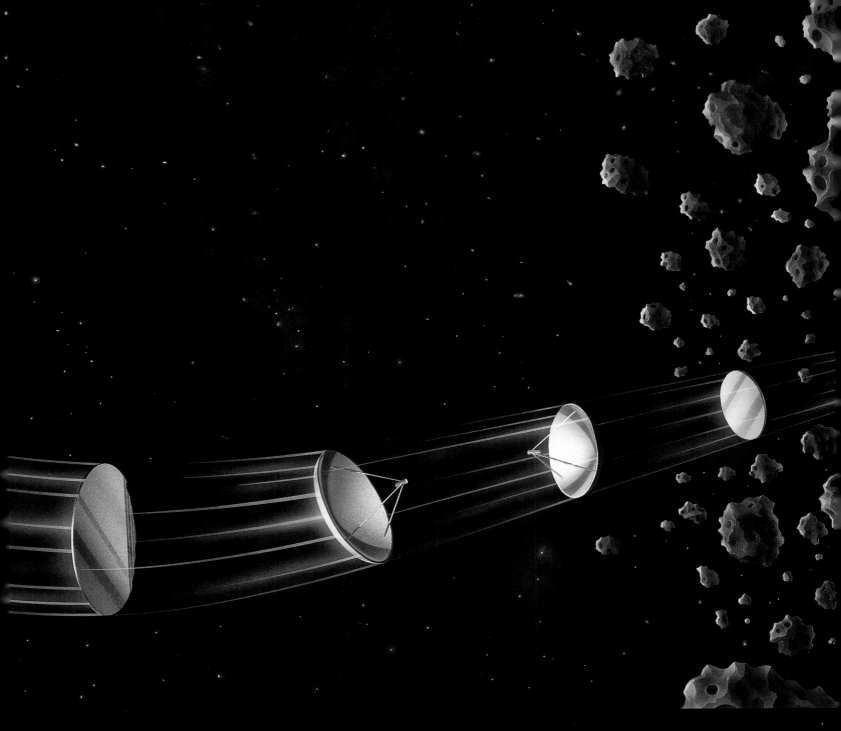

ergy. Shortly after departure, therefore, the on-board army of nanomachines begin to transform the craft into a microscopically thin and very reflective solar sail ten feet in diameter *(above, center).* Photons of sunlight caroming off the sail push the probe through space. To make a course correction, the nanomachines reshape the sail, crumpling one side accordion-style so that the uneven number of photons striking either side of the vessel's center of mass provide a turning force. The celestial tacking completed, the sail re-

sumes its shape and its journey. Two years after launch, with the probe approaching asteroids in the orbit of Jupiter, the nanomachines convert the sail into a parabolic optical telescope *(above, right)* that seeks out suitable landing spots. The preferred landfall will be an asteroid about twelve miles in diameter and rich in carbon, the element most easily digested by the nanomachines *(overleaf).* The probe then performs an about-face to transmit its list of candidates to Earth for final selection.

NANOMACHINES IN ACTION

Once a suitable asteroid has been selected, the probe crash-lands on its surface *(above)* and disgorges its cargo of nanomachines. Obeying their internal instructions, the machines use the asteroid's carbon atoms as raw material to replicate themselves. When the nanomachines exist in sufficient numbers, they begin to fashion a telescope dish, spreading the asteroid's aluminum atoms into a gossamer-thin parabolic mirror some two million miles in diameter *(above, center)*. Constructed atom by atom, and incorporating nanomachines that continue to adjust its shape, the completed dish is free of the imperfections caused by such crude methods as grinding, pounding, and polishing.

Next, the tiny assemblers make a tripod to sup-

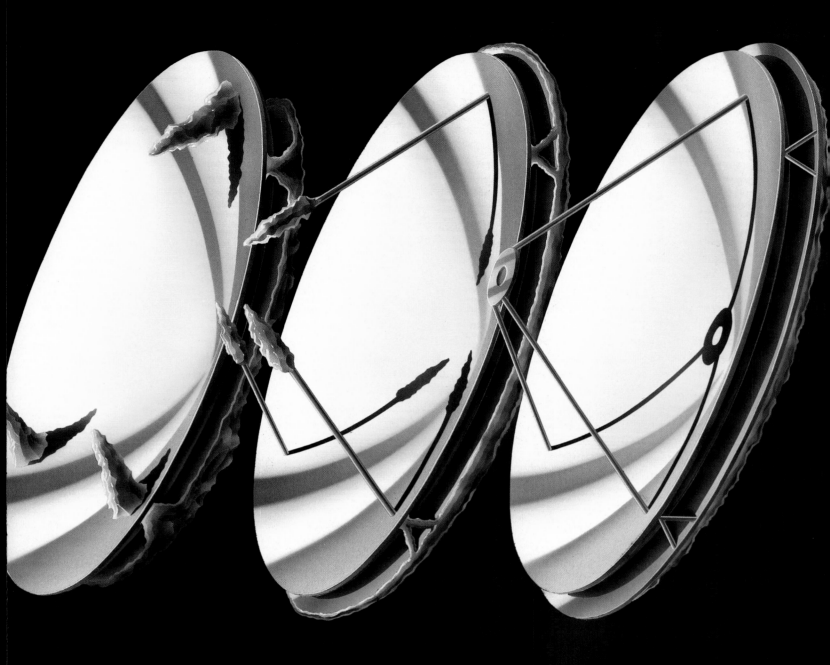

port the dish by turning carbon atoms into three diamond-fiber struts that meet at the mirror's focal point. When this tripod is in place, the machines transform some of the asteroid's remaining matter into imagers and spectrometers that will collect visible, infrared, and ultraviolet light. The nanomachines then fabricate a device called a transceiver to forward the information so gathered to controllers on Earth.

To collect solar energy to power the telescope's instruments and transceiver, the nanomachines cover the back of the dish with a sunshade that also shields the infrared detectors from the heat that would garble their readings. Finally, the nanomachines build small electromagnetic guns around the rim of the telescope. By firing minuscule pellets made of icy waste, the guns can swivel the dish into any observing position. Given a twelve-mile-wide host asteroid, the vastly larger finished telescope may be able to distinguish features as much as a hundred feet wide on planets ten light-years away.

in themselves, providing all the physical and psychological necessities to maintain a healthy human colony.

Conceivably, by the time suitable engines are developed, appropriate vessels may lie ready and waiting in the form of space colonies stationed near the Earth. The huge orbiting habitats first envisioned by Gerard O'Neill of Princeton in the late 1960s were not originally intended to voyage, but if they were retrofitted with fusion engines, perhaps, or gigantic laser sails, they would make crackerjack interstellar ships, capable of transporting an entire society of cosmic explorers.

Another possibility is J. D. Bernal's 1929 notion of a hollowed-out asteroid star cruiser—an idea resurrected and refined in 1964 by American author and space program analyst Dandridge Cole. In his scheme, space-based mirrors would focus the Sun's rays to create a giant cutting torch that would bore a hole through the center of a relatively small asteroid. The hole would be filled with tanks of water and the entire asteroid heated by the mirrors until it reached a molten state. Small rockets would then set the asteroid spinning to whip it into a cylindrical shape. When the heat reached the center, the water tanks would explode under steam pressure, blowing up the asteroid like a balloon and forming a vast cavity that could be as much as twenty miles long and ten miles in diameter. The cooled crust of the asteroid would protect the inhabitants from the dangers of cosmic radiation and would also shield them from space debris during the long interstellar missions.

Any kind of colony starship faces its share of problems, however. Simply building one, for example, would cost on the order of hundreds of billions of dollars—a staggering sum that could be provided only through the cooperative efforts of the world's richest nations. And because it could weigh as much as a billion tons, such a ship would be so slow that flights would last centuries. Finding thousands of people willing to sever all their earthly connections, knowing they would not even win the age-old explorer's reward of reaching the goal, would be a challenge in itself.

A less expen-

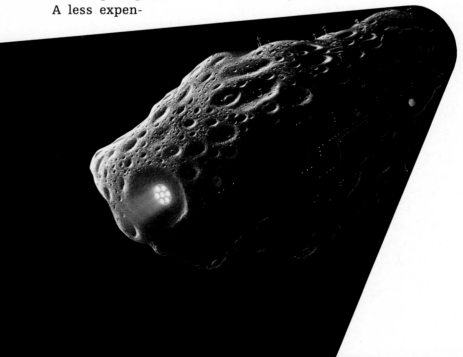

In 1929, British physicist J. D. Bernal published *The World, the Flesh and the Devil,* in which he suggested that spacefarers could live within a hollowed-out asteroid, using its metals and ices as construction materials. As other writers embellished on his description of the ship as a self-contained world, it became a prototype for fictional interstellar arks, home to generations of star travelers.

sive alternative to sending thousands of active, supply-consuming voyager-colonists to the stars would be to dispatch a smaller crew of explorers that had been put into hibernation, the "deep sleep" imagined by Goddard. Granted, sleeping passengers would require a computer system of unprecedented sophistication and reliability to monitor their vital functions and awaken them upon arrival, but such a crew would have minimal needs in terms of nourishment and living space. The ship could thus be light and swift, capable of realizing the maximum speeds of state-of-the-art propulsion systems. Still, merely slowing down biological functions, though sufficient for relatively short voyages—to the handful of stars within ten light-years of Earth, say—would probably be inadequate for expeditions to more distant corners of the galaxy; those trips might require stopping the biological clock altogether by freezing the crew.

Science-fiction writers have been enamored of this idea for decades, but it may be the most problematic of all the proposals. Cryobiology—the study of the effects of extremely low temperatures on living organisms—has turned up some sobering facts. Experiments have demonstrated that tissues and organs freeze and thaw at different rates, depending on their composition and their location in the body. During the thawing process, they would buckle and tear, much like sidewalks and streets during winter. Also, researchers have found that as water within individual cells freezes into ice crystals, it expands and ruptures cell walls, causing extensive damage. Some types of cells might be able to regenerate after thawing, but brain cells never regenerate and so would be irretrievably lost.

Although freezing an entire person may prove impossible, techniques for freezing embryos have already been developed and have suggested to some an even more radical solution. Given sufficient advances in computer technology, it might be possible to program a system that would thaw such embryos at a suitable point in a deep-space journey and bring them to term. The infant crew members would be reared and educated by computers stocked with the wisdom of terrestrial civilization. But such a scenario raises some disturbing questions. What would machine-reared humans be like? Who would decide when science had crossed the line between education and indoctrination? And what about the rights of these engineered individuals to choose their own destiny?

UPPING THE SPEED STAKES

Regardless of its practicality or ethical acceptability, this approach remains fundamentally unsatisfying, as does the idea of multigenerational starships. The stars will not seem truly reachable until they can be brought within the grasp of a single human lifetime. Time and again, then, proponents of starfaring reexamine ways of going faster and getting there sooner. Although many in the scientific community doubt that schemes for more exotic forms of travel will ever bear fruit, master science-fiction author and space-travel aficionado Arthur C. Clarke has pointed out that predictions of technological advances tend to be overoptimistic in the short run and underoptimistic in the long run. The British Interplanetary Society's Daedalus study, for instance, had the apparent virtue of being limited to technologies already in existence or likely to be perfected within the next fifty years or so—but that may have been its chief defect. On a longer time scale, ideas that now seem farfetched may prove to be more practical than the society's comparatively conservative plans.

ANNIHILATION

One concept that has attracted increasing attention since the late 1950s is an interstellar engine driven by the mutual annihilation of matter and antimatter. First posited by theoretical physicist Paul Dirac in the 1920s, antimatter is not quite as bizarre as it sounds. Every so-called elementary particle of matter, such as the electron or proton, has an antimatter counterpart, which has the same mass but is opposite in other characteristics, such as charge or spin. The key feature for propulsion researchers is that when two such particles come into contact, they totally destroy each other, their entire mass converting into energy. In fission and fusion reactions, by contrast, only a small percentage of matter becomes energy. An engine that could harness the energy released in matter-antimatter annihilations would be incredibly powerful, producing an Isp of up to several million seconds and a probable speed of 60 psol. At that rate, a journey from Earth to some of the nearest stars might involve a round trip of less than twenty-five years.

The champion of matter-antimatter (M-AM) technology is Robert Forward, a California-based research scientist instantly recognizable by his thick shock of silver hair, a penchant for wearing loud, paisley-patterned vests, and

an unflagging enthusiasm for futuristic voyaging. In the early 1980s, Forward received a grant from the Rocket Propulsion Laboratory at Edwards Air Force Base in southern California to study alternative propulsion energy sources. He took a year's leave from Hughes Research Labs to investigate, among other things, laser sailing and his chief passion, matter-antimatter drives.

As a part-time writer of science-fiction novels, Forward is not put off by what seem to be the insurmountable difficulties of dealing with antimatter—including finding a way to create and then store enough antimatter for a long voyage. Antimatter can be created in nuclear particle accelerators by aiming a beam of protons at a target atom, which disintegrates in a shower of elementary particles. Researchers are now experimenting with methods for segregating and storing the antimatter particles, using a powerful electromagnetic containment ring to prevent the antimatter from coming in contact with ordinary matter. But the process requires huge amounts of energy. By one estimate, if the entire energy output of the United States were devoted to the production of antimatter, the annual harvest would be a mere 300 pounds, far short of the several tons that would probably be needed for an interstellar rocket. And storage time has so far been limited to a few days at most.

Since Forward fully expects that science will one day solve these problems, he keeps his sights on the eventual capabilities of an M-AM starship, whose passengers would reap an unusual though entirely real benefit from the laws of relativity. As an object approaches the speed of light, the flow of time—from the object's perspective—actually slows down, a fact that has been verified experimentally; subatomic particles that normally decay very quickly have been shown to survive for longer periods when they are accelerated to very fast, or relativistic, velocities. The effect is small at speeds below 50 psol, but aboard an M-AM ship traveling at 60 psol, a trip lasting ten years as measured from Earth would be only eight years long for the crew. And if a way could be found to reach a speed as high as 99 psol, interstellar distances would shrink dramatically; the entire galaxy could be circumnavigated in twenty-five shipboard years.

SHORTCUTS

At about this point, discussions of interstellar travel begin to move into the twilight zone between science and science fiction. A good many astrophysicists, navigating the ever-surprising terrain of a relativistic cosmos, have dared to imagine strange shortcuts built into the very fabric of the universe. Einstein's special theory of relativity, which seems to preclude faster-than-light voyaging, actually only prohibits travel at the speed of light, at which point mass becomes infinite; the equations allow for the existence of particles on the other side of the light-speed barrier, dubbed tachyons by Columbia University physicist Gerald Feinberg from the Greek word for "fast." At present, tachyons are no more than mathematical abstractions, but their discovery—if it ever came to pass—would undoubtedly spur serious thought about ways to harness their time-warping speed.

The cosmic mysteries known as black holes could open an equally intriguing window in time. According to relativity theory, space and time together can be imagined as a sheet of rubber that is distorted and curved by the presence of mass. Black holes are thought to be objects of such incredible density that anything within reach of their gravitational pull, including light, cannot escape. Some black holes are so massive that they may even rip the fabric of space-time, creating "wormholes" that might provide passage to distant cosmic neighborhoods. By analogy, an ant crawling across the surface of an inflatable globe would have to walk a fair distance to get from Chicago to Shanghai. But if the globe were deflated and crumpled up—its spatial dimensions distorted, as if by a black hole—then the ant might find that

Theoretical physicist Robert Forward has long been one of the leading thinkers in the field of futuristic propulsion systems, investigating the possibilities of spaceships driven by a photon "wind" of laser light and starship engines fueled by antimatter.

Chicago and Shanghai are actually touching each other, and a simple step off the surface of the globe would complete the journey. This step out of normal space, through what some refer to as hyperspace, might make possible virtually instantaneous interstellar travel.

As utterly fantastic and even preposterous as such concepts seem for the moment, visionaries are encouraged by the inherent unpredictability of scientific breakthroughs and revelations. How and when humans first travel to the stars will depend on whether the big breakthroughs are in propulsion technology, biomedical science, or some totally unanticipated field. As Arthur Clarke has said, "Any sufficiently advanced technology is indistinguishable from magic."

The whirlwind of speculation that constitutes planning for galactic travel is proof of one indisputable truth: The human spirit cannot tolerate the unreachable. The yearning to explore is part of the biology of the species and, given enough time, will doubtless overcome a myriad of obstacles. The best perspective may be the long view that casts starfaring in an evolutionary light. University of Hawaii anthropologist Ben Finney and Eric Jones, a space scientist at the Los Alamos National Laboratory, have attempted to condense all of human history into four major developments. The first step occurred four million years ago when apelike prehumans came down from the trees and occupied the African grasslands. The second step was the migration from Africa into the rest of the world. Step number three began about 3,000 years ago, when people learned how to sail the seas. Step number four, say Finney and Jones, is the biggest one of all but as inevitable as the rest—the voyage from Earth to the stars.

IN SEARCH OF NEW SUNS

Science fiction often represents a journey between worlds as a kind of relativistic detour around three-dimensional space, a "warp-speed" leap that traverses light-years in an instant. The hard truth is that ships powered by today's chemical propulsion methods would take a minimum of 50,000 years to cross from the Sun's domain to that of the nearest star. But far greater speeds are almost certainly attainable. Ships of the future may carry humankind through the cosmos by means of one of the three propulsion techniques described on the following pages: beamed energy, nuclear fusion, and matter-antimatter annihilation.

All three are based on accepted scientific principles, but none is practical today. Beamed energy requires lasers far more powerful and accurate than any currently available; at present, controlled fusion uses more energy than it releases; and antimatter is extraordinarily difficult to produce and store. Despite these obstacles, many scientists believe that travel into deep space is no mere dream. After all, flight velocity already has increased more than 400,000 percent since the biplanes of 1905.

127

RIDING A WAVE OF BEAMED ENERGY

A laser beam that might be used in spaceship propulsion begins in a gas-filled chamber bounded by a fully reflective mirror *(right)* and a partially transmitting mirror *(far right)*. As shown here in simplified form, when an electron in the gas is electrically excited, it emits a photon, which strikes another electron, releasing a photon of the same wavelength and phase. After bouncing between the mirrors, identical photons emerge as a narrow, high-energy beam; corrected for diffraction, it can travel millions of miles without significant spreading.

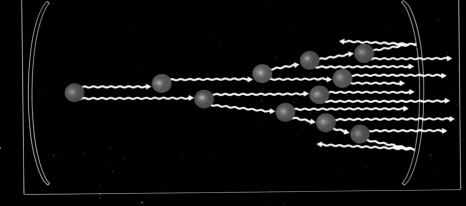

Shielded by a 300-foot-wide plate, a ship rides toward deep space on a powerful beam of laser light. A satellite in low Earth orbit (far left) sends low-power light to a station on Earth, which amplifies it and sends it back to the satellite. Relayed to the spacecraft, the beam heats hydrogen propellant to accelerate the ship to a small percentage of the speed of light.

THE POWER OF NUCLEAR FUSION

In a hypothetical fusion engine, a doughnut-shaped torus (shown in part) contains a gas of deuterium (one proton and one neutron) and helium-3 (two protons and one neutron), confined by a powerful magnetic field *(green ribbon).* Heated to more than 900 million degrees Fahrenheit and then injected with more gas, the atoms of deuterium and helium-3 fuse into an unstable atom of lithium-5, which instantly breaks into helium-4 and one proton. A tiny amount of the mass of lithium-5 is then converted to immense amounts of energy.

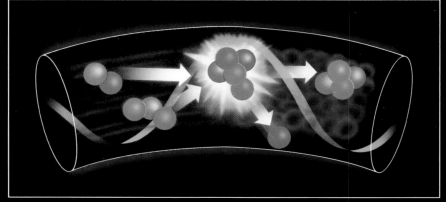

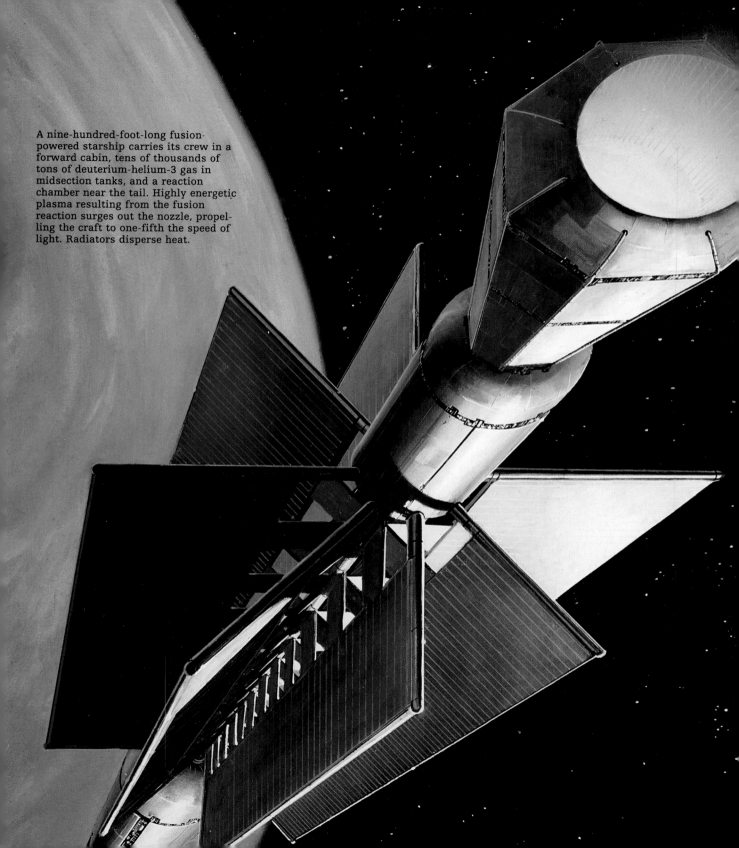

A nine-hundred-foot-long fusion-powered starship carries its crew in a forward cabin, tens of thousands of tons of deuterium-helium-3 gas in midsection tanks, and a reaction chamber near the tail. Highly energetic plasma resulting from the fusion reaction surges out the nozzle, propelling the craft to one-fifth the speed of light. Radiators disperse heat.

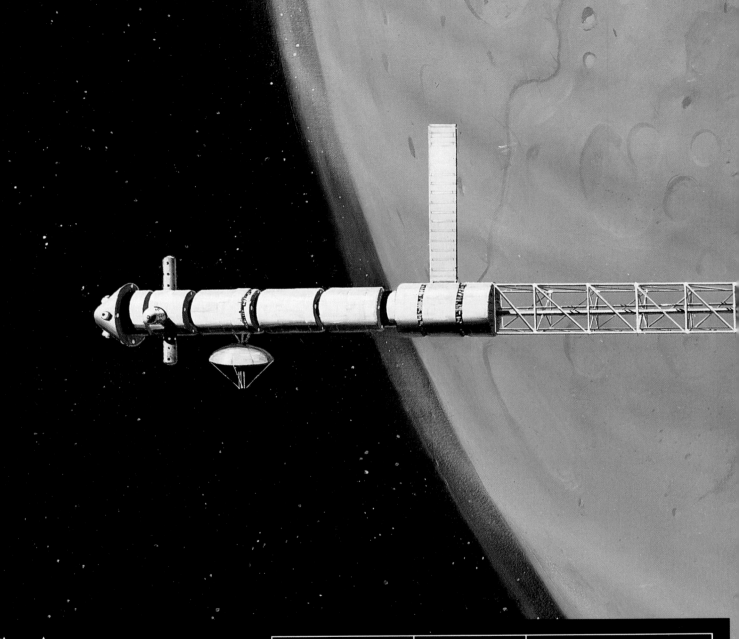

An Annihilation Engine

In a matter-antimatter engine, a frozen crystal of antihydrogen is levitated within an electromagnetic field *(near right)*. Ultraviolet light knocks free a stream of antiprotons, and magnetic coils guide the antiparticles toward the nozzle. When the beam meets a stream of regular hydrogen *(green spheres)*, the two annihilate each other, creating energetic subatomic particles and gamma rays. The particles rush out the nozzle at 94 percent light-speed, flinging the craft in the opposite direction and eventually accelerating it close to the speed of light.

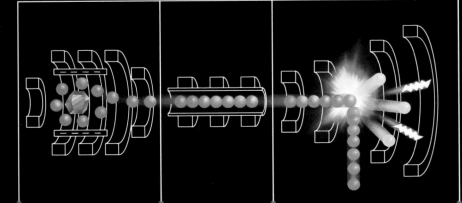

A ship powered by matter-antimatter would be very long to put as much distance as possible between the crew quarters *(above, far left)* and deadly gamma rays produced in the annihilation chamber behind the three hydrogen tanks *(above, right)*. Radiation shields protect the magnets and other sensitive ship components, and radiators remove excess heat from the reaction chamber.

GLOSSARY

Accelerator: a device that senses change in speed along an axis.

Accretion disk: a disk formed from gases and other materials drawn in by a compact body, such as a black hole, at the disk's center.

Aerobraking: a procedure for decelerating a spacecraft upon its arrival at a planet; instead of firing retrorockets, the craft makes a shallow pass through the planet's atmosphere, using the resulting friction to slow down.

Aeroponics: the cultivation of plants without soil by suspending them on wire shelves and feeding their roots with a nutrient mist.

Air lock: a compartment that separates areas of different environment, especially different air pressures; used for entry to and departure from a spacecraft or extraterrestrial habitat.

Altimeter: a device that measures altitude above the surface of a planet or moon. Spacecraft altimeters work by timing the round trip of radio signals bounced off the surface.

Antimatter: matter made up of antiparticles identical in mass to matter particles but opposite in such properties as electrical charge. For example, a positively charged positron is the antiparticle to a negatively charged electron.

Asteroid: a small, rocky, airless body that orbits a star.

Asteroid belt: a zone between the orbits of Mars and Jupiter occupied by several thousand asteroids.

Attitude: a spacecraft's orientation with respect to its direction of motion.

Basalt: a dark, crystallized molten rock that most commonly solidifies on the surface in the form of lava flows. It is the dominant rock type in the lunar maria.

Beryllium: a metallic element with low density and a high melting point that may be used to make protective shields for spacecraft.

Black hole: in theory, an extremely compact body with such great gravitational force that no electromagnetic radiation can escape from it.

Booster: a rocket used to launch spacecraft.

Boron: an element that could be used to reinforce metals for the building of microthin sails on interstellar spacecraft.

Cherenkov radiation: radiation emitted by a particle moving through a medium faster than the speed of light in that medium. High-energy cosmic rays produce Cherenkov radiation, visible as faint blue light, when they strike Earth's atmosphere.

Conjunction: an apparent meeting or close approach of two celestial bodies as viewed from Earth. In planetary astronomy, if only one planet is mentioned, the second body in the conjunction is the Sun. Inferior conjunction occurs when Venus or Mercury passes between the Sun and Earth; superior conjunction occurs when a planet is directly behind the Sun.

Cosmic ray: an atomic nucleus or other charged particle moving at close to the speed of light.

Deuterium: a form of hydrogen having one neutron and one proton in its nucleus. Also known as heavy hydrogen.

Electromagnetism: the force that attracts oppositely charged particles and repels similarly charged particles. Electromagnetism does not affect neutral particles such as neutrinos.

Electron: a negatively charged particle that normally orbits an atom's nucleus but may exist in isolation.

Escape velocity: the minimum speed needed for an object's momentum to carry it out of the gravitational pull of a massive object such as a planet or moon.

Fission: a nuclear process that releases energy when heavyweight nuclei break down into lighter nuclei.

Fusion: a nuclear process that releases energy when lightweight atomic nuclei combine to form a heavier nucleus.

Gamma rays: the most energetic form of electromagnetic radiation, with the highest frequency and shortest wavelength of all forms of the spectrum.

Geosynchronous: describing the orbit of a spacecraft or satellite that completes a circle every twenty-four hours, the same time Earth requires to make one rotation; thus the object remains above one location on the ground. Geosynchronous orbits are established over the equator at an altitude of approximately 22,000 miles. Also called geostationary.

Gravity: the force responsible for the mutual attraction of separate masses. *See also* Microgravity and Zero gravity.

Hohmann transfer orbit: a trajectory in which one end of a spacecraft's elliptical path touches the orbit of Earth and the other end intersects the orbit of the destination planet or moon. Because it takes advantage of gravity, a spacecraft traveling a Hohmann transfer trajectory requires less fuel than for other flight paths.

Hydrazine: a corrosive, reducing liquid base composed of two parts nitrogen and four parts hydrogen; used in rocket fuels.

Hydroponics: the cultivation of plants by placing the roots in an inert medium and feeding them with a liquid nutrient solution.

Ilmenite: a mineral compound made up of iron, oxygen, and titanium.

Infrared: a band of electromagnetic radiation with a lower frequency and a longer wavelength than visible red light.

Intragalactic: within the boundaries of the Milky Way galaxy.

Ionosphere: an atmospheric layer dominated by charged, or ionized, atoms that extends from about 38 to 400 miles above Earth's surface.

Laser (from "light amplification by stimulated emission of radiation"): a device that produces a narrow beam of high-intensity monochromatic radiation at infrared, optical, or shorter wavelengths.

Libration point: one of five locations between Earth and its moon where the gravitational pull of the two bodies is balanced.

Light-year: an astronomical distance unit equal to the distance light travels in a vacuum in one year, almost six trillion miles.

Mare, maria: from the Latin for "sea," low plains of the Moon.

Maser (from "microwave amplification by stimulated emission of radiation"): a device or celestial object that excites molecules to produce a narrow beam of radio waves of a certain wavelength.

Mass driver: an electromagnetic catapult for launching objects into space.

Matter-antimatter annihilation (M-AM): a process in which equal amounts of matter and antimatter collide and destroy each other, producing a burst of energy that could propel a spacecraft.

Meteoroid: a small metallic or rocky body that is found

in space. A meteoroid entering a planet's atmosphere is called a meteor. Meteors often burn up in the atmosphere; those that reach the surface of the planet are known as meteorites.

Microgravity: an environment—within an orbiting spacecraft, for example—of very weak gravitational forces. Microgravity conditions in space stations may allow experiments or manufacturing processes that are not possible on Earth.

Micrometeor: a meteor the size of a grain of dust.

Neutrino: a chargeless particle with little or no mass, given off during nuclear fusion.

Neutron: a chargeless particle with a mass similar to a proton's; normally found in an atom's nucleus.

Oort cloud: a cloud of comets orbiting the Sun at a distance of 30,000 to 100,000 astronomical units. (An astronomical unit is the distance between the Sun and Earth—93,000,000 miles.)

Opposition: the alignment of two celestial bodies on opposite sides of the sky as seen from Earth. Oppositions occur when Earth passes between a planet and the Sun; a perfect opposition occurs when a planet is also at its closest approach to Earth.

Orbit: the path of an object revolving around another object; also, to progress along such a periodic path.

Parking orbit: a temporary orbit for a spacecraft or satellite, established to rendezvous with other spacecraft, or to wait for the right time to transfer to an advantageous trajectory.

Photon: a unit of electromagnetic energy associated with a specific wavelength. It behaves as a chargeless particle traveling at the speed of light.

Plasma: a gaslike association of ionized particles that responds collectively to electric and magnetic fields.

Positron: an antimatter particle similar in mass to an electron, but carrying a positive electric charge.

Propellant: a chemical or chemical mixture burned to create thrust for a rocket or spacecraft.

Proton: a positively charged particle with about 2,000 times the mass of an electron; normally found in the nucleus of an atom.

Radio interferometer: an instrument for examining sources of radio waves through the simultaneous use of two or more separated telescopes. Interferometers produce overlapping wave patterns from the radiation; the patterns are studied to determine the wavelength and angular diameter of the emitting source.

Radio telescope: an instrument for studying astronomical objects at radio wavelengths.

Radio waves: the least energetic form of electromagnetic radiation, with the lowest frequency and the longest wavelength.

Ramjet: a jet engine that uses external propellant sources, such as gas from the interstellar medium.

Regolith: unconsolidated fragmental rock debris that covers the surface of the Moon; also called lunar soil.

Resolution: the degree to which details in an image can be separated, or resolved. The resolving power of a telescope is usually proportional to the diameter of its mirror or aperture.

Retrorocket: a rocket that produces thrust in a direction opposite to the direction of motion to slow a larger rocket or satellite.

Rocket: a missile or vehicle propelled by the combustion of a fuel and a contained oxygen supply. The forward thrust of a rocket results when exhaust products are ejected from the tail.

Solar panel: an array of light-sensitive cells that can be attached to a spacecraft and used to generate electrical power for the vehicle in space.

Solid rocket booster: a rocket, powered by solid propellants, used to launch spacecraft into orbit.

Space habitat: an orbiting spacecraft designed to support human activity for an extended period.

Spectrometer: an instrument that splits light or other electromagnetic radiation into its individual wavelengths, or spectrum, and records the results electronically.

Spectrum: the array of colors or wavelengths obtained by dispersing light from a star or another source, as through a prism. Spectra are often striped with emission or absorption lines, which indicate the composition and can show the motion of the light source.

Speed of light: 186,000 miles per second.

Superconductivity: the propensity of certain solid materials to conduct electric current without resistance or loss of energy when greatly cooled.

Superluminal: appearing to travel faster than the speed of light.

Supersonic: moving faster than the speed of sound.

Tachyon: a theoretical particle that can travel faster than the speed of light, perhaps a billion times faster.

Trajectory: the curve traced by an object moving through space. A closed trajectory is an orbit.

Transponder: a device that transmits a response signal automatically when activated by an incoming signal.

Ultraviolet: a band of electromagnetic radiation that has a higher frequency and a shorter wavelength than visible blue light.

X-rays: a band of electromagnetic radiation intermediate in wavelength between ultraviolet radiation and gamma rays.

Zero gravity: a condition in which gravity appears to be absent. Zero gravity occurs when gravitational forces are balanced by the acceleration of a body in orbit or free fall.

BIBLIOGRAPHY

Books

Adams, John R., *Edward Everett Hale*. Boston: G. K. Hall, 1977.

Adelman, Saul J., and Benjamin Adelman, *Bound for the Stars*. Englewood Cliffs, N.J.: Prentice-Hall, 1981.

Angelo, J. A., Jr., *The Extraterrestrial Encyclopedia*. New York: Facts On File, 1985.

Belew, Leland F., ed., *Skylab: Our First Space Station*. Washington, D.C.: NASA Scientific and Technical Information Office, 1977.

Bergaust, Erik, *Wernher von Braun*. Washington, D.C.: National Space Institute, 1976.

Bernal, J. D., *The World, the Flesh and the Devil.* London: Jonathan Cape, 1970.

Boston, Penelope J.:

"Critical Life Science Issues for a Mars Base." In *The Case for Mars II.* Vol. 62 of Science and Technology Series, ed. by Christopher P. McKay. San Diego, Calif.: Univelt, 1984.

"Mars Mission Life Support." In *The NASA Mars Conference.* Vol. 71 of Science and Technology Series, ed. by Duke B. Reiber. San Diego, Calif.: Univelt, 1988.

Calder, Nigel:

Einstein's Universe. New York: Viking, 1979.

Spaceships of the Mind. London: British Broadcasting Corporation, 1978.

Carr, Michael H., *The Surface of Mars.* New Haven and London: Yale University Press, 1981.

Caudill, Thomas R., "Mass-Balance Model for a Controlled Ecological Life Support System on Mars." In *The Case for Mars II.* Vol. 62 of Science and Technology Series, ed. by Christopher P. McKay. San Diego, Calif.: Univelt, 1984.

Chappell, Russell E., *Apollo.* Washington, D.C.: National Geographic Society, 1973.

Clark, Phillip, *The Soviet Manned Space Program.* New York: Crown, 1988.

Clarke, Arthur C.:

Arthur C. Clarke's July 20, 2019: Life in the 21st Century. New York: Macmillan, 1986.

The Exploration of Space. New York: Harper & Brothers, 1951.

Considine, Donald M., ed., *Van Nostrand's Scientific Encyclopedia.* New York: Van Nostrand Reinhold, 1983.

Cooper, Henry S. F., Jr., *A House in Space.* New York: Holt, Rinehart and Winston, 1976.

Cortright, E. M., ed., *Apollo Expeditions to the Moon.* Washington, D.C.: NASA Scientific and Technical Information Office, 1975.

Cowley, Stewart, *Spacecraft: 2000 to 2100 AD.* Secaucus, N.J.: Chartwell Books, 1978.

Drexler, K. Eric, *Engines of Creation.* New York: Doubleday, 1986.

Dyson, Freeman J., *Infinite in All Directions.* New York: Harper & Row, 1988.

Finke, Robert C., ed., *Electric Propulsion and Its Applications to Space Missions.* Vol. 79 of *Progress in Astronautics and Aeronautics.* New York: American Institute of Aeronautics and Astronautics, 1981.

Forward, Robert L., and Joel Davis, *Mirror Matter: Pioneering Antimatter Physics.* New York: John Wiley & Sons, 1988.

Freeman, Michael, *Space Traveller's Handbook.* New York: Sovereign Books, 1979.

French, James R.:

"Aerobraking and Aerocapture for Mars Missions." In *The Case for Mars.* Vol. 57 of Science and Technology Series, ed. by Penelope J. Boston. San Diego, Calif.: American Astronautical Society, 1981.

"The Impact of Martian Propellant Manufacturing on Early Manned Exploration." In *The Case for Mars II.* Vol. 62 of Science and Technology Series, ed. by Christopher P. McKay. San Diego, Calif.: Univelt, 1984.

Friedman, Louis, *Starsailing: Solar Sails and Interstellar Travel.* New York: John Wiley & Sons, 1988.

Gatland, Kenneth, *The Illustrated Encyclopedia of Space Technology.* New York: Crown, 1981.

Gernsback, Hugo, *Ultimate World.* New York: Walker, 1971.

Giancoli, Douglas C., *Physics: Principles with Applications.* Englewood Cliffs, N.J.: Prentice-Hall, 1985.

Glenn, Jerome Clayton, and George S. Robinson, *Space Trek: The Endless Migration.* Harrisburg, Pa.: Stackpole Books, 1978.

Grey, Jerry, *Enterprise.* New York: Morrow, 1979.

Gross, Robert A., *Fusion Energy.* New York: John Wiley & Sons, 1984.

Hale, Edward Everett, "The Brick Moon." In *Masterpieces of Science Fiction,* ed. by Sam Moskowitz. New York: World, 1966.

Harrison, Harry, and Malcolm Edwards, *Spacecraft in Fact and Fiction.* New York: Exeter Books, 1979.

Hartmann, William K., Ron Miller, and Pamela Lee, *Out of the Cradle: Exploring the Frontiers beyond Earth.* New York: Workman, 1984.

Heppenheimer, T. A.:

Colonies in Space. Harrisburg, Pa.: Stackpole Books, 1977.

Toward Distant Suns. Harrisburg, Pa.: Stackpole Books, 1979.

Kaufmann, William J., III, *Universe.* New York: W. H. Freeman, 1987.

Lehman, Milton, *This High Man: The Life of Robert H. Goddard.* New York: Farrar, Straus, 1963.

Lewis, John S., and Ruth A. Lewis, *Space Resources: Breaking the Bonds of Earth.* New York: Columbia University Press, 1987.

Life in Space, by the Editors of Time-Life Books. Alexandria, Va.: Time-Life Books, 1983.

Locher, Frances C., ed., *Contemporary Authors.* Detroit: Gale Research, no date.

McAleer, Neil, *The Omni Space Almanac: A Complete Guide to the Space Age.* New York: Pharos Books, 1987.

McDonough, Thomas R., *Space: The Next Twenty-Five Years.* New York: John Wiley & Sons, 1987.

McKay, Christopher P., "Living and Working on Mars." In *The NASA Mars Conference.* Vol. 71 of Science and Technology Series, ed. by Duke B. Reiber. San Diego, Calif.: Univelt, 1988.

Metzger, Linda, and Deborah A. Straub, eds., *Contemporary Authors.* Detroit: Gale Research, no date.

Miles, Frank, and Nicholas Booth, eds., *Race to Mars: The Mars Flight Atlas.* New York: Harper & Row, 1988.

Miller, Ron, and William K. Hartmann, *The Grand Tour: A Traveller's Guide to the Solar System.* New York: Workman, 1981.

Moore, Patrick:

The Moon. New York: Rand McNally, 1981.

Travellers in Space and Time. Garden City, N.Y.: Doubleday, 1983.

Moore, Patrick, ed., *The International Encyclopedia of Astronomy.* New York: Orion Books, 1987.

Moritz, Charles, ed., *Current Biography Yearbook 1980.* New York: H. W. Wilson, no date.

National Commission on Space, *Pioneering the Space Frontier.* New York: Bantam Books, 1986.

Nicholls, Peter, ed., *The Science in Science Fiction.* New York: Crescent Books, 1982.

Nicholson, Iain, *The Road to the Stars.* Devon, U.K.: Westbridge Books, 1978.

Nicogossian, Arnauld E., "Human Factors for Mars Missions." In *The NASA Mars Conference.* Vol. 71 of Science and Technology Series, ed. by Duke B. Reiber. San Diego, Calif.: Univelt, 1988.

Norris, Geoffrey, "Konstantin E. Tsiolkovsky." In *Rocket and Missile Technology,* ed. by Lt. Col. Gene Gurney, U.S.A.F. New York: Franklin Watts, 1964.

Oberg, James E.:
Mission to Mars: Plans and Concepts for the First Manned Landing. Harrisburg, Pa.: Stackpole Books, 1982.
Red Star in Orbit. New York: Random House, 1981.

Oberg, James E., and Alcestis R. Oberg, *Pioneering Space: Living on the Next Frontier.* New York: McGraw-Hill, 1986.

O'Neill, Gerard K., *The High Frontier: Human Colonies in Space.* Garden City, N.Y.: Anchor Press, 1982.

Outbound (Voyage through the Universe series). Alexandria, Va.: Time-Life Books, 1989.

Pauling, Linus, *General Chemistry.* San Francisco: W. H. Freeman, 1959.

Powers, Robert M., *The Coattails of God.* New York: Warner, 1981.

Sagan, Carl, *Cosmos.* New York: Random House, 1980.

Shatalov, V. A., M. F. Rebrov, and E. A. Vaskevich, *To the Stars.* Moscow: Planeta Publishers, 1986.

Shipman, Harry L., *Space 2000: Meeting the Challenge of a New Era.* New York: Plenum Press, 1987.

Siegel, Mark, *Hugo Gernsback, Father of Modern Science Fiction.* San Bernardino, Calif.: Borgo Press, 1988.

Ste. Croix, Philip de, ed., *The Illustrated Encyclopedia of Space Technology.* New York: Crown, 1981.

Stoker, C. R., et al., "Scientific Program for a Mars Base." In *The Case for Mars II.* Vol. 62 of Science and Technology series, ed. by Christopher P. McKay. San Diego, Calif.: Univelt, 1984.

Taggart, Arthur F., *Handbook of Mineral Dressing.* New York: John Wiley & Sons, 1945.

Tibbits, T. W., and D. K. Alford, eds., *Controlled Ecological Life Support System: Use of Higher Plants.* Washington, D.C.: NASA Scientific and Technical Information Office, 1982.

Tsiolkovsky, Konstantin, *The Call of the Cosmos.* Moscow: Foreign Languages Publishing House, no date.

Vogt, Gregory, *An Album of Modern Spaceships.* New York: Franklin Watts, 1987.

Von Braun, Wernher, *Space Frontier.* New York: Holt, Rinehart and Winston, 1971.

Von Braun, Wernher, and Frederick I. Ordway III, *Space Travel: An Update of History of Rocketry & Space Travel.* New York: Harper & Row, 1985.

Welch, Steven, "Mission Strategy and Spacecraft Design for a Mars Base Program." *The Case for Mars II.* Vol. 62 of Science and Technology series, ed. by Christopher P. McKay. San Diego, Calif.: Univelt, 1984.

White, Frank, *The Overview Effect: Space Exploration and Human Evolution.* Boston: Houghton Mifflin, 1987.

Periodicals

Asimov, Isaac, "The Next Frontier?" *National Geographic,* July 1976.

Baker, Peter, "The Search for a Lunar Oasis." *Final Frontier,* December 1988.

Biryukov, Yuri, "Stairway to the Stars." *Soviet Life,* April 1989.

Boslough, John, "Searching for the Secrets of Gravity." *National Geographic,* May 1989.

Boston, Penelope J., "Low-Pressure Greenhouses and Plants for a Manned Research Station on Mars." *Journal of the British Interplanetary Society,* May 1981.

Bova, Ben, "Full (Moon) Employment." *Space World,* March 1988.

Bulban, Erwin J., "Economic Benefits of Lunar Base Cited." *Aviation Week & Space Technology,* April 18, 1983.

Burke, James D., "Living on the Moon." *Planetary Report,* March-April 1985.

Burnham, Robert, "How Apollo Changed the Moon." *Astronomy,* July 1989.

"Bush Proposed Establishment of a U.S. Moon Base and Exploration of Mars." *Wall Street Journal,* July 21, 1989.

Bussard, Robert, "A Starship is Born—the Latest Designs." *Science Digest,* May 1983.

Cassenti, B. N., "A Comparison of Interstellar Propulsion Methods." *Journal of the British Interplanetary Society,* March 1982.

Chaikin, Andrew:
"Moonrocks." *Life,* July 1989.
"Return to the Moon." *Sky & Telescope,* June 1983.

Chernow, Ron, "Colonies in Space May Turn Out to be Nice Places to Live." *Smithsonian,* February 1976.

Chernyshov, Mikhail, "We've Got the Cosmos in Our Hands." *Air & Space,* April-May, 1989.

Chien, Philip, "Assembling the Space Station." *Space World,* December 1987.

Collins, Michael, "Mission to Mars." *National Geographic,* November 1988.

"Colonizing Space." *Time,* May 26, 1975.

Cook, William J., "The New Frontiers." *U.S. News & World Report,* September 26, 1988.

Covault, Craig, "Manned U.S. Lunar Station Wins Support." *Aviation Week & Space Technology,* November 19, 1984.

Cox, Randal, "Zero-G Mysteries and LifeSat." *Space World,* March 1988.

Craxton, R. Stephen, Robert L. McCrory, and John M. Soures, "Progress in Laser Fusion." *Scientific American,* August 1986.

"Daedalus: A New View." *Spaceflight,* May 1985.

Daniloff, Nicholas, "The Space Statesman." *Air & Space,* October-November 1988.

David, Leonard, "Mars: The Next Giant Leap?" *Final Frontier,* April 1988.

Dewdney, A. K., "Computer Recreations." *Scientific American,* January 1988.

Dooling, David, Jr., "Controlled Thermonuclear Fusion for Space Propulsion." *Spaceflight,* January 1972.

Dorr, Les, Jr., "The Future As It Was." *Final Frontier,* July-August 1989.

Duke, Michael, "The Role of a Lunar Base in Mars Exploration." *Planetary Report,* March-April 1988.

Farquhar, Robert W.:
"Detour to a Comet: Journey of the International

Cometary Explorer." *Planetary Report,* May-June 1985.
"Future Missions for Libration-Point Satellites."
Astronautics & Aeronautics, May 1969.
"A Halo-Orbit Lunar Station." *Astronautics & Aeronautics,* June 1972.
Forward, Robert, "A Programme for Interstellar Exploration." *Journal of the British Interplanetary Society,* October 1976.
Franklin, Kenneth L.:
"Gravitational Forces and Effects." *Natural History,* October 1963.
"Gravitational Forces and Effects Part II." *Natural History,* November 1963.
Frisbee, Robert H., "Propulsion Systems." *Space Education,* October 1985.
Gatland, K. W., "Project Orion." *Spaceflight,* December 1974.
Gelman, David, et al., "Colonies in Space." *Newsweek,* November 27, 1978.
Henson, Carolyn, " '77 NASA Summer Study: No Show Stoppers!' *Space World,* December 1977.
Heppenheimer, T. A., "Small World." *Final Frontier,* March-April, 1989.
"Interview: Gerard O'Neill." *Final Frontier,* April 1988.
Kerr, R. A., "Making the Moon, Remaking Earth." *Science,* March 17, 1989.
Kitt, Michael T., "Observe the Apollo Landing Sites." *Astronomy,* July 1989.
Kolm, Henry, "An Electromagnetic 'Slingshot' for Space Propulsion." *Space World,* February 1978.
Kolm, Henry, J. Oberteuffer, and David Kelland, "High Gradient Magnetic Separation." *Scientific American,* November 1975.
Kolm, Henry, and Richard D. Thornton, "Electromagnetic Flight." *Scientific American,* October 1973.
Koltz, Charles, "Mining the Moon." *Space World,* February 1983.
Lemonick, Michael D., "The Next Giant Leap for Mankind." *Time,* July 24, 1989.
Lewis, Richard S., "Space Prospect: Factories and Electric Power." *Smithsonian,* December 1977.
Loftus, John P., "Man's Role in Space Exploration and Exploitation." *Spaceflight,* June 1987.
McCurdy, Howard E., "The Graying of Space." *Space World,* October 1988.
McKay, Christopher, "The Case for Mars." *Planetary Report,* March-April 1985.
Mallove, Eugene F., "Renaissance in the Search for Galactic Civilizations." *Technology Review,* January 1984.
Maranto, Gina, "Earth's First Visitors to Mars." *Discover,* May 1987.
Martin, A. R.:
"Mankind's Interstellar Future." *Spaceflight,* February 1984.
"Some Limitations of the Interstellar Ramjet." *Spaceflight,* January 1972.
Maryniak, Gregg, "The Bounty of Space." *Space World,* August-September 1979.
Matloff, Gregory L., "Utilization of O'Neill's Model 1 Lagrange Point Colony as an Interstellar Ark." *Journal of the British Interplanetary Society,* December 1976.
Michaud, M. A. G., and L. E. David, "Return to the Moon."

Astronomy, April 1980.
Morgan, Joseph W., "Superrelativistic Interstellar Flight." *Spaceflight,* July 1973.
Moser, Thomas L., "Space Station Freedom Getting Ready to Go." *Aerospace America,* September 1988.
Mueller, George, "Antimatter and Distant Space Flight." *Spaceflight,* May 1983.
Nichols, Robert G., "From Footprints to Foothold." *Astronomy,* July 1989.
Nolley, Betty, "Dining in the Stars." *Space World,* November 1988.
O'Leary, Brian, "The Space Studies Institute." *Space World,* April 1978.
O'Neill, Gerard, K.:
"The Colonization of Space." *Physics Today,* September 1974.
"Engineering a Space Manufacturing Center." *Astronautics & Aeronautics,* October 1976.
"The New American Frontier." *Discover,* November 1985.
"Progress toward Space Manufacturing." *Space World,* January 1977.
"Space Colonies and Energy Supply to the Earth." *Science,* December 5, 1975.
"Out of This World." *Fortune,* June 1974.
Powers, Robert:
"Fast Forward: A Conversation with Robert L. Forward." *Space World,* January 1987.
"Propulsion Future." *Space World,* September 1985.
"Project Daedalus: An Interim Report on the BIS Starship Study." *Spaceflight,* September 1974.
Reichhardt, Tony, "Roads to Mars." *Final Frontier,* March-April 1989.
Rein, Richard K., "Maybe We Are Alone—Physicist Gerry O'Neill Says That's a Reason for Sending People into the Safety of Space." *People,* December 12, 1977.
Reinert, Al, "Gerry's World." *Air & Space,* April-May 1989.
Sawyer, Kathy, "Orbiting Lab Raises Questions of Mission, Need." *The Washington Post,* April 20, 1989.
Schefter, Jim, "On Base in Space." *Popular Science,* March 1989.
Shevchenko, V. V., "A Soviet View of a Lunar Base." *Planetary Report,* November-December 1988.
Smith, Harlan J., "Astronomy from the Moon." *Sky & Telescope,* July 1987.
"Sociology on the Space Station: An Interview with B. J. Bluth." *Space World,* January 1986.
"Space Update: Lunar Observatory." *Space World,* September 1984.
"Space World Special Report: America's Return to the Moon." *Space World,* July 1984.
Spangenburg, Ray, and Diane Moser, "The Loneliest Place on Earth." *Final Frontier,* April 1988.
"Special Report: The Next Ten Years in Space." *Final Frontier,* January-February 1989.
Steinhart, Peter, "Age of the Mesocosm." *Sierra,* September-October 1988.
Sternbach, Rick, "Journey to the Stars." *Science Digest,* May 1983.
Stewart, Doug, "High Society." *Air & Space,* October-November 1988.
Stoker, Carol, and Christopher P. McKay, "Mission to

Mars: The Case for a Settlement." *Technology Review,* November-December, 1983.

Strong, James, "Further Thoughts on Interstellar Exploration." *Spaceflight,* May 1981.

"A Talk with John Aaron." *Planetary Report,* March-April 1988.

Taylor, Richard L. S., "Manned Spaceflight: The Human Barrier." *Space,* January-February 1988.

Taylor, Stuart Ross, "The Origin of the Moon." *American Scientist,* September-October 1987.

"They Slipped the Surly Bonds of Earth to Touch the Face of God." *Time,* February 10, 1986.

"Trying to Live a Year in Space." *Soviet Life,* July 1989.

Von Braun, Wernher:
"Can We Get to Mars?" *Collier's,* April 30, 1954.
"Crossing the Last Frontier." *Collier's,* March 22, 1952.

Wagstaff, Bill, "A Spaceship Named Orion." *Air & Space,* October-November 1988.

Wolsky, Alan M., Robert F. Giese, and Edward J. Daniels, "The New Superconductors: Prospects for Applications." *Scientific American,* February 1989.

Other Sources

"Advanced Propulsion Systems Concepts for Oribital Transfer." NASA Catalog of Advanced Propulsion Concepts D180-26680-1. Washington, D.C.: NASA, no date.

Alred, J., et al., "Development of a Lunar Outpost: Year 2000-2005." Paper No. LBS-88-240, Lunar Bases and Space Activities in the 21st Century Conference, NASA Johnson Space Center, Houston, Tex., April 5-7, 1988.

Archard, Karen R., "Shuttle Based Assembly of Space Station." Report to NASA. Seal Beach, Calif.: Rockwell International Space Transportation Systems Division, 1988.

"Assembly Sequence CRBB000468 Supporting Data." *Space Station Freedom Program Level II System Engineering and Integration.* Publication No. SSE-E-88-R20. Washington, D.C.: NASA, September 23, 1988.

Billingham, John, William Gilbreath, and Brian O'Leary, eds., "Space Resources and Space Settlements." Technical papers for the 1977 Summer Study at NASA Ames Research Center, Moffett Field, Calif. NASA Paper No. SP-428. Washington, D.C.: NASA Scientific and Technical Information Branch, 1979.

Bluth, B. J., "Lunar Settlements: A Socio-economic Outlook." Ref. No. IAF-86-513, 37th Congress of the International Astronautical Federation, Innsbruck, Austria, October 4-11, 1986.

Bluth, B. J., and Martha Helppie, "Soviet Space Stations as Analogs" (2nd ed.). Washington, D.C.: NASA, August 1986.

Burns, Jack O.:
"Some Astronomical Challenges for the 21st Century." Paper No. LBS-88-025, Lunar Bases and Space Activities in the 21st Century Conference, NASA Johnson Space Center, Houston, Tex., April 5-7, 1988.
"21st Century Astronomical Observatories on the Moon." Reprinted from *Engineering and Construction, and Operations in Space.* Albuquerque, N.M.: Aerospace Division/ASCE, August 29-31, 1988.

Burns, Jack O., et al., "Astronomical Observatories on the Moon." Unpublished paper. Available from author,

University of New Mexico, Department of Physics and Astronomy, Albuquerque, N.M. 87131, no date.

Burns, Jack O., and Wendell W. Mendell, eds., "Future Astronomical Observatories on the Moon." NASA Conference Publication 2489. Washington, D.C.: NASA Scientific and Technical Information Division, 1988.

Burns, Jack O., et al., eds., Proceedings of the Workshop on a Lunar Far-Side Very Low Frequency Array, held at the BDM Corporation, Albuquerque, N.M., February 18-19, 1988.

Bush, Harold G., et al., "Conceptual Design of a Mobile Remote Manipulator System." NASA Technical Memorandum 86262. Hampton, Va.: NASA Langley Research Center, July 1984.

"The Case for Mars: Concept Development for a Mars Research Station." JPL Publication No. 86-28. Pasadena, Calif.: NASA Jet Propulsion Laboratory, April 15, 1986.

"Controlled Ecological Life Support System." Pamphlet prepared by the Biomedical Operations and Research Office at the John F. Kennedy Space Center, Florida, no date.

David, Leonard, "Space Station Freedom: A Foothold on the Future." Washington, D.C.: NASA Office of Space Station, no date.

Duke, Michael B., and John W. Alred, "A Lunar Base Scenario Emphasizing Early Self-Sufficiency." Paper No. LBS-88-241, Lunar Bases and Space Activities in the 21st Century Conference, NASA Johnson Space Center, Houston, Tex., April 5-7 1988.

Eagle Engineering, Inc.:
"Advanced Space Transportation System Support Contract Summary Final Report." NASA Contract NAS9-17878. Houston, Tex.: Eagle Engineering, October 30, 1988.
"Conceptual Design of a Lunar Oxygen Pilot Plant." EEI Report 88-182, NASA Contract No. NAS9-17878. Houston, Tex.: Eagle Engineering, July 1988.
"Lunar Surface Operations Study." EEI Report 87-172, NASA Contract No. NAS9-17878. Houston, Tex.: Eagle Engineering, December 1987.

"EASE/ACCESS Postmission Management Report: EVA 1: Demonstrating Construction in Orbit." Huntsville, Ala.: NASA Marshall Space Flight Center, no date.

"Exploring the Living Universe: A Strategy for Space Life Sciences." Report of the NASA Life Sciences Strategic Planning Study Committee. Washington, D.C.: NASA, June 1988.

Fairchild, Kyle, and Wendell W. Mendell, "Report of the In Situ Resources Utilization Workshop." NASA Conference Publication 3017. Washington, D.C.: NASA Scientific and Technical Information Division, 1988.

Forward, Robert L., "Making and Storing Antihydrogen for Propulsion." In *Low Energy Antimatter,* proceedings of the workshop on the design of a low energy antimatter facility held at the University of Wisconsin-Madison, October 1985. World Scientific, 1986.

"Fusion Power." Information Bulletin NT-1. Princeton, N.J.: Princeton University Plasma Physics Laboratory, no date.

Gibson, Michael A., and Christian W. Knudsen:
"Lunar Oxygen Production via Fluidized Bed Iron Oxide Reduction." Paper for presentation at American Insti-

tute of Chemical Engineers Symposium on Non-
Traditional Applications of Chemical Reaction Engineering, Washington, D.C., November 27-December 2,
1988.

"Lunar Oxygen Production from Ilmenite." Houston, Tex.:
Carbotek Inc., 1984.

Greene, W. H., and R. T. Haftka, "Reducing Distortion and
Internal Forces in Truss Structures by Member Ex-
changes." NASA Technical Memorandum 101535.
Hampton, Va.: NASA Langley Research Center, January
1989.

Heard, Walter L., Jr., et al., "Astronaut/EVA Construction
of Space Station." Hampton, Va.: NASA Langley Research Center, 1988.

"Information Summaries: Space Station." NASA Report
No. PMS-008A (Hqs). Washington, D.C.: NASA, August
1988.

Johnson, Richard D., ed., "Space Settlements: A Design
Study." NASA Report No. SP-413. Washington, D.C.:
NASA Scientific and Technical Information Office, 1977.

Johnson, Stewart W., et al., "Developing Concepts for
Lunar Astronomical Observatories: Interdisciplinary
Team Experiences at the University of New Mexico."
Paper for presentation at 1989 ASEE Annual Confer-
ence, Lincoln, Nebr., June 25-29, 1989.

"Life Sciences Report." Washington, D.C.: NASA Life
Sciences Division Office of Space Science and Applications, December 1987.

MacGillivray, Charles Scott, "Design and Development of
the Truss Assembly Fixture for Space Station Assembly
Operations." Report to NASA. Seal Beach, Calif.:
Rockwell International Satellite and Space Electronics
Division, 1988.

"Manned Space Station Attached Payload Accomodations
Handbook." NASA Announcement of Opportunity No.
OSSA-1-88. Greenbelt, Md.: Goddard Space Flight
Center, January 19, 1988.

Maryniak, Gregg E., "First Steps to Lunar Manufacturing:
Results of the 1988 Space Studies Institute Lunar
Systems Workshop." Princeton, N.J.: Space Studies
Institute, 1988.

Mikulas, Martin M., Jr., and Harold G. Bush, "Design,
Construction and Utilization of a Space Station Assem-
bled from 5-Meter Erectable Struts." NASA Technical
Memorandum 89043. Hampton, Va.: NASA Langley
Research Center, October 1986.

Mikulas, Martin M., Jr., R. C. Davis, and W. H. Greene, "A
Space Crane Concept: Preliminary Design and Static
Analysis." NASA Technical Memorandum 101498.
Hampton, Va.: NASA Langley Research Center, Novem-
ber 1988.

NASA Office of Aeronautics and Space Technology:
"Project Pathfinder." Report No. 400-330. Washington,
D.C.: NASA, 1987.

NASA Office of Exploration:
"Beyond Earth's Boundaries: Human Exploration of the
Solar System in the 21st Century." Annual Report to the
NASA Administrator. Washington, D.C.: NASA, 1988.
"Exploration Studies Technical Report." Technical
Memorandum 4075. Washington, D.C.: NASA, December
1988.
"Office of Exploration Overview April 1989." Washing-
ton, D.C.: NASA, April 3, 1989.

Oder, R. R.:
"Engineering Aspects of Dry Magnetic Separation." In
Novel Concepts, Methods and Advanced Technology in
Particulate-Gas Separation, proceedings of workshop
held at the University of Notre Dame, Ind., April 1977.
"High Gradient Magnetic Separation Theory and
Applications." IEEE Transactions on Magnetics, Vol.
MAG-12, No. 5. New York: Institute of Electrical and
Electronics Engineers, September 1976.
"Magnetic Beneficiation of Lunar Ores." Report pre-
sented by EXPORTech Co. Inc., at EMEC Consultants
Workshop on Dry Extraction of Silicon and Aluminum
from Lunar Ores, Pittsburgh, Pa., November 7, 1988.

Oder, R. R., Lawrence A. Taylor, and Rudolf Keller,
"Magnetic Characterization of the Lunar Soils." In
Proceedings of Lunar & Planetary Science XX. Houston,
Tex.: March 1989.

"Pathfinder: In-Space Assembly and Construction Project
Plan." Hampton, Va.: NASA Langley Research Center,
December 1988.

Ride, Sally K., "Leadership and America's Future in
Space." A Report to the Administrator. Washington,
D.C.: NASA, August 1987.

Roberts, M. L., "Inflatable Habitation for the Lunar
Base." Paper No. LBS-88-266, Lunar Bases and Space
Activities in the 21st Century Conference, NASA
Johnson Space Center, Houston, Tex., April 5-7, 1988.

Schmid, P. E., "Lunar Far-Side Communication Satellites."
Report X-507-67-328. Greenbelt, Md.: Goddard Space
Flight Center, July 28, 1967.

Smith, Marcia S., "Space Activities of the United States,
Soviet Union, and Other Launching Countries: 1957-
1987." CRS Report for Congress. Washington, D.C.:
Library of Congress, Congressional Research Service,
February 29, 1988.

Snively, Leslie O., and Gerard K. O'Neill, "Mass Driver III:
Construction, Testing and Comparison to Computer
Simulation." Princeton, N.J.: Space Studies Institute, no
date.

"Space Station Freedom Reference Guide." Huntsville,
Ala.: Boeing, 1988.

"Space Station: A Research Laboratory in Space." NASA
Pamphlet PAM 512. Washington, D.C.: NASA, no date.

Space Studies Institute:
"SSI Update: Special Report on High Frontier
Research." Newsletter. Princeton, N.J.: Space Studies
Institute, 1986.
"SSI Update." Newsletter. Princeton, N.J.: Space Studies
Institute, September-October 1987.
"SSI Update." Newsletter. Princeton, N.J.: Space Studies
Institute, January-February 1988.

Stofan, Andrew J., "Space Station: A Step into the
Future." NASA Pamphlet PAM 510. Washington, D.C.:
NASA, November 1987.

Teren, Fred, "Space Station Electric Power System
Requirements and Design." NASA Technical Memorandum 89889. Washington, D.C.: NASA, 1987.

U.S. Senate, Committee on Commerce, Science, and
Transportation, "Soviet Space Programs: 1981-87."
Committee print. Washington, D.C.: U.S. Government
Printing Office, May 1988.

University of Wisconsin, "Astrofuel for the 21st Century."
Madison, Wis.: College of Engineering, March 1988.

INDEX

ACKNOWLEDGMENTS

The editors wish to thank Vickie J. Baker, NASA Langley Research Center, Hampton, Va.; Andrea Bardroff, Messerschmitt-Bölkow-Blohm, Munich; Sybille Boecker, Leiterin Öffentlichkeitsarbeit MBB, Munich; Susan W. Bonner, NASA Johnson Space Center, Houston, Tex.; Stanley Borowski, Dave Byers, NASA Lewis Research Center, Cleveland, Ohio; L. J. Carter, The British Interplanetary Society, London; Stewart Cobb, Redondo Beach, Calif.; Steve Cohen, NASA Lewis Research Center, Cleveland, Ohio; Mark Craig, NASA Johnson Space Center, Houston, Tex.; Leonard David, Space Data Resources and Information, Washington, D.C.; John T. Dorsey, NASA Langley Research Center, Hampton, Va.; Bob Durrett, Marshall Spaceflight Center, Huntsville, Ala.; Carter Emmart, Demarest, N.J.; European Space Agency, Washington, D.C.; Robert Forward, Forward Unlimited, Malibu, Calif.; Louis Friedman, The Planetary Society, Pasadena, Calif.; Robert Harrington, U.S. Naval Observatory, Washington, D.C.; Walter L. Heard, NASA Langley Research Center, Hampton, Va.; Mark Hess, NASA Headquarters, Washington, D.C.; Alan Ladwig, NASA Office of Exploration, Washington, D.C.; Greg Lange, NASA Johnson Space Center, Houston, Tex.; Agnès Letraublon Public Relations, CNES, Toulouse, France; David McKay, NASA Johnson Space Center, Houston, Tex.; Ian Mathers, TRI Design Studios, Southampton, U.K.; Harvey Myerson, U.S. International Space Year, Washington, D.C.; John Niehoff, Science Applications International Corporation, Schaumburg, Ill.; Robin Oder, EXPORTech Co., Export, Pa.; Carl Pilcher, NASA Headquarters, Washington, D.C; Udo Pollvogt, Messerschmitt-Bölkow-Blohm, Arlington, Va.; Jacques Simon, Responsable Système, CNES, Toulouse, France; Marcia Smith, Library of Congress, Washington, D.C.; John Soldner, NASA Johnson Space Center, Houston, Tex.; Kathy Hoyt, United States Geologic Survey, Flagstaff, Ariz.; Robert Williams, Defense Advanced Research Projects Agency, Arlington, Va.; Brian Wowk, Winnipeg, Manitoba, Canada.

PICTURE CREDITS

The sources for the illustrations in this book are listed below. Credits from left to right are separated by semicolons; credits from top to bottom are separated by dashes.

Cover: Art by Bryn Barnard, photographed by Larry Sherer. 2-9: Art by Paul Hudson. 14, 15: NASA, LBJ Space Center, Houston, Tex. (61D-45-0046). 16: Initial cap, detail from pages 14, 15. 19: Library of Congress Archives; from *Space Travel* by Wernher von Braun and Frederick I. Ordway III, Harper & Row, New York, 1985, copied by Renée Comet. 20, 21: Soviet Life; from *To the Stars* by V. A. Shatalov, M. F. Rebrov, and E. A. Vaskevich, Planeta Publishers, Moscow, 1986, copied by Renée Comet; from *Amazing Stories,* Teck Publications, Chicago, 1933, copied by Renée Comet; photo copied by Kirk Hammond, courtesy Forrest Ackerman; from *Amazing Stories,* copied by Renée Comet. 22, 23: From *The Exploration of Space* by Arthur C. Clarke, Harper & Row, New York, 1951, courtesy British Planetary Society, London; Pictorial Press, London; Frederic W. Freeman Trust; photo by James Minnich, Fairchild Industries, Inc.; from *The Journal of the British Interplanetary Society,* London, 1951, copied by Renée Comet; Kelly/Mooney Photography. 24: NASA, Washington, D.C. (74-HC-7); Tass/Sovfoto. 26, 27: NASA, Washington, D.C. (89-HC-12). 29: Art by Fred Holz. 30-33: Art by Stephen Wagner. 35: Larry Sherer, courtesy NASA, Langley Research Center, Hampton, Va.—Art by Fred Holz. 36-47: Art by Joe Bergeron, photographed by Larry Sherer, inset art by Fred Holz. 48, 49: NASA, LBJ Space Center, Houston, Tex. (AS17-142-21801); (AS17-142-21805); (AS17-142-21807); (AS-17-142-21811). 50: Initial cap, detail from pages 48, 49. 52: NASA, Washington, D.C. (72-HC-922). 55: James A. Bryant/© 1987 Discover Publications. 57-61: Art by Stephen Wagner. 62, 63: Art by Fred Holz. 65: Eagle Engineering, Inc., Houston, Tex. 66: NASA, Washington, D.C. 69: Art by Fred Holz. 70-75: Art by Rob Wood and Yvonne Gensurowsky of Stansbury, Ronsaville, Wood, Inc. 76, 77: USGS, Flagstaff, Ariz. 78: Initial cap, detail from pages 76, 77. 80: NASA, LBJ Space Center, Houston, Tex. (S85-41284). 82: USGS, Flagstaff, Ariz. 83: NASA, Washington, D.C. (89-HC-76). 85: © 1989 Roger Ressmeyer/Starlight (2). 86: NASA, Washington, D.C. (87-HC-174). 89-93 Art by Stephen Bauer. 94: Tass/Sovfoto. 100-104: Art by Rob Wood and Yvonne Gensurowsky of Stansbury, Ronsaville, Wood, Inc., based on original art by Carter Emmart. 106, 107: Geoff Chester. 108: Initial cap, detail from pages 106, 107. 110: Lee E. Battaglia; Christopher Springmann. 112: Model and photo by Matt Irvine. 114: Salamander Books, Ltd. 117-121: Art by Stephen Wagner. 122: David A. Hardy. 124: Cliff Olson. 127: Art by Bryn Barnard. 128-133: Art by Bryn Barnard, photographed by Larry Sherer, inset art by Fred Holz.

Time-Life Books
is a wholly owned subsidiary of
THE TIME INC. BOOK COMPANY

President and Chief Executive Officer:
Kelso F. Sutton
President, Time Inc. Books Direct:
Christopher T. Linen

TIME-LIFE BOOKS INC.
EDITOR: George Constable
Executive Editor: Ellen Phillips
Director of Design: Louis Klein
Director of Editorial Resources: Phyllis K. Wise
Editorial Board: Russell B. Adams, Jr., Dale M.
Brown, Roberta Conlan, Thomas H. Flaherty, Lee
Hassig, Jim Hicks, Donia Ann Steele, Rosalind
Stubenberg
Director of Photography and Research:
John Conrad Weiser

PRESIDENT: John M. Fahey, Jr.
Senior Vice Presidents: Robert M. DeSena, James
L. Mercer, Paul R. Stewart, Curtis G. Viebranz,
Joseph J. Ward
Vice Presidents: Stephen L. Bair, Bonita L.
Boezeman, Stephen L. Goldstein, Juanita T.
James, Andrew P. Kaplan, Trevor Lunn, Susan J.
Maruyama, Robert H. Smith
Supervisor of Quality Control: James King

PUBLISHER: Joseph J. Ward

Editorial Operations
Copy Chief: Diane Ullius
Production: Celia Beattie
Library: Louise D. Forstall

Computer Composition: Gordon E. Buck
(Manager), Deborah G. Tait, Monika D. Thayer,
Janet Barnes Syring, Lillian Daniels

Correspondents: Elisabeth Kraemer-Singh (Bonn),
Christina Lieberman (New York), Maria Vincenza
Aloisi (Paris), Ann Natanson (Rome).

VOYAGE THROUGH THE UNIVERSE

SERIES DIRECTOR: Roberta Conlan
Series Administrator: Judith W. Shanks

Editorial Staff for *Spacefarers*
Designer: Dale Pollekoff
Associate Editor: Sally Collins (pictures)
Text Editors: Pat Daniels (principal),
Allan Fallow, Lee Hassig
Researchers: Katya Sharpe Cooke, Tina McDow-
ell, Edward O. Marshall
Writer: Robert M. S. Somerville
Assistant Designer: Brook Mowrey
Copy Coordinators: Anne Farr, Darcie Conner
Johnston
Picture Coordinator: Ruth Moss
Editorial Assistant: Katie Mahaffey

Special Contributors: Amy Aldrich, J. Kelly
Beatty, Gregory Byrne, Jim Dawson, Jane Farrell,
Tom Heppenheimer, Alan MacRobert, Dennis
Overbye, Peter Pocock, Eugene Rodgers, Chuck
Smith, John Sullivan, Mark Washburn (text);
Adam Dennis, Mark Galan, Mark Lazen, Christina
Moore, Philip Murphy, Jacqueline Shaffer,
Eugenia Scharf, Marilyn Murphy Terrell
(research); Elizabeth Graham (index).

CONSULTANTS
PENELOPE J. BOSTON, a microbiologist and atmospheric chemist, is director of research for Complex Systems Research, Inc., a nonprofit organization dedicated to the space sciences.

JACK O. BURNS is director of the Institute for As-
trophysics at the University of New Mexico. He is
also a consultant in space plasma physics at the Los
Alamos National Laboratory.

BEN CLARK, an aerospace engineer and space scientist, works at Martin Marietta, Littleton, Colorado, where he specializes in Mars mission studies.

STEPHEN COOK works on space station configuration at NASA Headquarters, Washington, D.C.,
where he is a policy analyst.

K. ERIC DREXLER is a visiting scholar at Stanford
University and an authority on nanotechnology, a
theoretical process that would use molecular machines to build objects to complex chemical specifications.

MICHAEL DUKE is chief of the Solar System Exploration Division at the Lyndon B. Johnson Space Center in Houston.

DAVID DUNHAM, an astronomer and aerospace engineer, works for Computer Sciences Corporation in
Lanham-Seabrook, Maryland, calculating orbits
and trajectories for spacecraft.

ROBERT FARQUHAR plans new space mission programs in space physics and solar system exploration. He works at Goddard Spaceflight Center,
Greenbelt, Maryland, and at NASA Headquarters in
Washington, D.C.

JAMES R. FRENCH, JR., an aerospace engineer in
La Canada, California, consults for government and
private clients on rocket propulsion, spacecraft design, and emission design for piloted and unpiloted
missions.

VICTORIA GARSHNEK specializes in space medicine at NASA Headquarters. She teaches at the
Space Policy Institute at George Washington University in Washington, D.C.

BRAND NORMAN GRIFFIN, an aerospace designer,
currently manages the habitation module for the
Space Station Program at Boeing Aerospace, Huntsville, Alabama.

CHRISTIAN W. KNUDSEN, a chemical engineer, is
president of Carbotek Inc., which has developed a
process to extract oxygen from lunar rock for use as
a propellant.

MICHAEL R. LA POINTE is an advanced propulsion
engineer with Sverdrup Technology, Inc., at the
NASA Lewis Research Center in Cleveland, Ohio.

URIEL LOVELACE has planned mission studies to
the Moon and Mars. Currently he is at NASA Langley
Research Center in Hampton, Virginia, studying advanced spacecraft concepts.

GREGG MARYNIAK is executive vice-president of
the Spaces Studies Institute in Princeton, New Jersey, a nonprofit organization that pioneered the use
of resources found in space.

MARTIN M. MIKULAS is the director of the Structural Mechanics Division at the NASA Langley Research Center.

LESLIE O. SNIVELY, a technical consultant to the
Space Studies Institute, designed and built Mass
Driver III. The device was a prototype for launching
material from the surface of the moon.

PATRICK A. TROUTMAN is an engineer at the NASA
Langley Research Center in the Space Station Freedom Office.

**Library of Congress Cataloging in
Publication Data**
Spacefarers/by the editors of Time-Life Books.
p.   cm.   (Voyage through the universe).
Bibliography: p.
Includes index.
ISBN 0-8094-6891-3.
ISBN 0-8094-6892-1 (lib. bdg.).
1. Outer space—Exploration. 2. Manned
spaceflight.
I. Time-Life Books.   II. Series.
TL793.S66 1990
629.45—dc20   89-4482 CIP

For information on and a full description of
any of the Time-Life Books series, please call
1-800-621-7026 or write:
Reader Information
Time-Life Customer Service
P.O. Box C-32068
Richmond, Virginia 23261-2068